SpringerBriefs in Complexity

AF762805

Springer Complexity

Springer Complexity is an interdisciplinary program publishing the best research and academic-level teaching on both fundamental and applied aspects of complex systems—cutting across all traditional disciplines of the natural and life sciences, engineering, economics, medicine, neuroscience, social and computer science.

Complex Systems are systems that comprise many interacting parts with the ability to generate a new quality of macroscopic collective behavior the manifestations of which are the spontaneous formation of distinctive temporal, spatial or functional structures. Models of such systems can be successfully mapped onto quite diverse "real-life" situations like the climate, the coherent emission of light from lasers, chemical reaction-diffusion systems, biological cellular networks, the dynamics of stock markets and of the internet, earthquake statistics and prediction, freeway traffic, the human brain, or the formation of opinions in social systems, to name just some of the popular applications.

Although their scope and methodologies overlap somewhat, one can distinguish the following main concepts and tools: self-organization, nonlinear dynamics, synergetics, turbulence, dynamical systems, catastrophes, instabilities, stochastic processes, chaos, graphs and networks, cellular automata, adaptive systems, genetic algorithms and computational intelligence.

The three major book publication platforms of the Springer Complexity program are the monograph series "Understanding Complex Systems" focusing on the various applications of complexity, the "Springer Series in Synergetics", which is devoted to the quantitative theoretical and methodological foundations, and the "SpringerBriefs in Complexity" which are concise and topical working reports, case-studies, surveys, essays and lecture notes of relevance to the field. In addition to the books in these two core series, the program also incorporates individual titles ranging from textbooks to major reference works.

More information about this series at http://www.springer.com/series/8907

Carlos M. Lemos

Agent-Based Modeling of Social Conflict

From Mechanisms to Complex Behavior

Springer

Carlos M. Lemos
Department of Religion, Philosophy
and History
University of Agder
Kristiansand, Norway

ISSN 2191-5326 ISSN 2191-5334 (electronic)
SpringerBriefs in Complexity
ISBN 978-3-319-67049-2 ISBN 978-3-319-67050-8 (eBook)
https://doi.org/10.1007/978-3-319-67050-8

Library of Congress Control Number: 2017952602

Printed on acid-free paper

This Springer imprint is published by Springer Nature
The registered company is Springer International Publishing AG
The registered company address is: Gewerbestrasse 11, 6330 Cham, Switzerland

To my wife São and my daughters
Mariana and Marta

Preface

Simulation of social conflict phenomena using agent-based models and the methods of complex system studies is a topic of growing interest, for it provides a degree of insight and understanding not attainable using classical methods of disciplines like anthropology, sociology, social psychology, and political science.

This book describes a study of large-scale conflict against a central authority using an extension of Epstein's agent-based model of civil violence, which started with and builds on my PhD dissertation. The key idea was to extend Epstein's classical model by including mechanisms such as dependence of grievance on relative deprivation, legitimacy feedback, and network influence effects, using simple formulations inspired on social conflict theories.

The scope of the work is the description of large-scale and low-intensity conflict phenomena that involve a significant proportion of the population and are (mostly) self-organized. The purpose of the study was to show how the newly introduced mechanisms increase the generative capacity of the original model and to discuss the plausibility of the patterns of magnitude, duration, and interval of the simulated conflict events, by comparing them with corresponding statistical descriptions of conflict events in some countries affected by the "Arab Spring." As in Epstein's model, the dynamics of conflict are described by the interaction between two populations of artificial agents: "citizens" which remain quiet or rebel and law-enforcing agents (or "cops") which arrest rebellious "citizens." Armed conflict and other conflict manifestations that involve higher violence intensity and organized structures (e.g. insurgences) are not considered. Religious and ethnic conflicts, which would require agents endowed with identity and consideration of more complicated micro-interactions, are also not studied.

The book is organized as follows. The first three chapters contain the introduction (Chap. 1), a summary review of social conflict theories and related concepts (Chap. 2), and a discussion of Epstein's model with emphasis on key variables and mechanisms (Chap. 3). Chapter 4 contains a statistical description of conflict events based on the Social Conflict Analysis Database and an analysis of indicators related to legitimacy, human rights, and inequality, for eight African countries affected by the "Arab Spring." This chapter addresses aspects of general interest, such as the usefulness and limitations of the information in databases of conflict events; the limited value of indices of legitimacy, human rights, and inequality as prognostic tools; and above all the complexity of real conflict processes. These analyses provide useful information for discussing the plausibility of computer simulations and illustrate the issues and limitations inherent to the parameterization and validation of agent-based models of social conflict. Chapter 5 contains a description of the agent-based model developed in this work, with emphasis on the extensions to Epstein's original model: vanishing of the risk perception below a critical ratio (set via an input parameter) between deterrence and "group support," deprivation-dependent hardship with variable sensitivity to deprivation (to represent the difference between political and economic deprivation), endogenous legitimacy feedback, and network influence effects modeled via dispositional contagion. Chapter 6 describes a set of simulation experiments that illustrate the generative capacity of the model and the influence of the newly introduced mechanisms on the complexity of the solutions. The results in this chapter highlight interesting aspects, such as how patterns of magnitude, duration, and recurrence of simulated events are influenced by input parameters, the occurrence of solutions with different regimes (calm, intermittent peaks of turmoil and permanent rebellion) in scenarios of low legitimacy and high repression, and the importance of deterrence vis-à-vis sensitivity to deprivation.

The present work will be of general interest to researchers working in social simulation using agent-based models. It will be of particular interest to those working on social conflict, under the perspectives of modeling (abstraction), the development and testing of theories, the use of indicators and databases, and the interplay between theory, models, and data analyses. Prospects for future work include improved modeling of network influences and extension of the model to religious and ethnic conflicts.

Kristiansand, Norway — Carlos M. Lemos
July 2017

Acknowledgments

I wish to express my gratitude to Professors Rui J. Lopes of the University Institute of Lisbon (ISCTE) and Helder Coelho of the Faculty of Sciences of the University of Lisbon, my former PhD advisors, for their encouragement and support.

Thanks are also due to Lieutenant-General Xavier Matias of the Portuguese Army who, as director of the Portuguese Joint Command and Staff College, always held the highest consideration and enthusiasm toward my work and gave high priority to funding my participation in conferences to present papers while I was a PhD student. Support by the BioISI (Instituto de Biossistemas e Ciências Integrativas) and by the Faculty of Sciences of the University of Lisbon during my PhD work is also gratefully acknowledged.

I also want to thank Professor Bruce Gilley of Portland State University, for sharing his data on legitimacy and presenting valuable explanations and suggestions for modeling legitimacy feedback.

In July 15th, 2016, I started working as a postdoctoral researcher in the Modeling Religion in Norway (MODRN) project, in the Department of Religion, Philosophy and History at the University of Agder, Kristiansand, Norway. This position provided excellent conditions and the computer resources I needed for exploring the model.

Therefore, I am grateful to the Research Council of Norway, grant number 250449, for funding my contract, and to all the staff of the Department of Religion, Philosophy and History at the University at Agder, for the collaboration and support I received since I started working in MODRN. In particular, I want to thank Professor F. LeRon Shults for his interest in my work, personal support, enthusiasm, and encouragement. I also thank Senior Engineer Sigurd Kristian Brinch of the Department of Information and Communication Technology at the University of Agder, for providing me access to computer resources that were vital for performing the simulation experiments and for his help on working via remote access.

I must also acknowledge the excellent support I received from Springer during the preparation of this manuscript for publication. I am grateful to editors Christian Caron and Christopher Coughlin for their interest in my work and personal support to its publication. I also want to thank assistant editor Ho Ying Fan and project coordinator Dhanuj Nair for all their fast yet patient and accurate replies to my queries.

Finally, I thank my wife São and my daughters Mariana and Marta for their continuous love, support, trust, and enthusiasm, without which this work would have never been completed. They were, are, and will always be my strongest source of energy and inspiration.

Contents

List of Symbols and Acronyms

α	Shape parameter in the Pareto distribution
α_i	Learning constant for the i-th agent in Rescorla-Wagner conditioning model (salience of conditioned stimulus)
β_i	Learning constant for the i-th agent in Rescorla-Wagner conditioning model (salience of unconditioned stimulus)
δ	Exponent in generalized Rescorla-Wagner conditioning model
$\mathscr{N}(\mu, \sigma^2)$	Normal distribution with mean value μ and variance σ^2
$\mathscr{U}(a, b)$	Continuous uniform distribution with support $[a, b]$
$\mathscr{A}$	Generic agent
$\mathscr{P}$	Agent's percept
$\mathscr{R}$	Set of possible finite runs
$\mathscr{S}$	Sum of social influences in the action rule for a "citizen" agent in the ABM of social conflict
$\mu(X)$	Mean (or expected) value of random variable X
ρ	C/A, the ratio between the number of "cops" and the number of "active" citizens in the ABM of social conflict
ρ_c	Critical "cop"-to-"active" ratio below which risk perception vanishes; a parameter in the proposed form of the estimated arrest probability function (P_a)
ρ_f	Fixed point value of the "cop"-to-"active" ratio in the analytical model described in Sect. 6.1
ρ_v	$= (C_v/A_v)$
$\mathtt{AG}_i$	The set of "active" citizens in the `group` network, in the ith "citizen" in the ABM of social conflict
$\mathtt{AINFL}_i$	The set of "active" citizens in the `infl` network, in the ith "citizen" in the ABM of social conflict
$\mathrm{var}(X)$	Variance of random variable X
$A(t)$	Total number of "active" citizens at time step t
A_v	Number of "active" citizens within a citizen's vision radius in the ABM of social conflict
Ac	Set of actions available to an agent $\mathscr{A}$

C	Number of "cops"
C_v	Number of "cops" within a citizen's vision radius in the ABM of social conflict
E	Set of possible environment states
G	Grievance
g_G	Probability density function of G
H	Hardship
J	Jail term for a citizen in the ABM of social conflict
$J(t)$	Total number of "jailed" citizens at time step t
$J_{\max}$	Maximum jail term in the ABM of social conflict
k	Arrest constant in the expression of the estimated arrest probability function (P_a)
L	Legitimacy
L_p	Citizens' perceived legitimacy (in the case of endogenous legitimacy feedback)
N	Net risk, $= R \cdot P_a$
n_{active}	Total number of "citizens" in the "active" state, in the ABM of Social conflict
$N_{citizen}$	Total number of citizens in the social conflict model
n_{jailed}	Total number of "citizens" in the "jailed" state, in the ABM of Social conflict
n_N	Probability density function of N
n_{quiet}	Total number of "citizens" in the "quiet" state, in the ABM of Social conflict
P	Deliberative component of disposition, in Epstein's *Agent_Zero* framework
R	Risk aversion
S	Social component of disposition, in Epstein's *Agent_Zero* framework
T	Threshold in threshold rules of binary decision
T_i	Threshold for agent $\mathscr{A}_i$
U_X	Probability distribution of X
u_X	Probability density function of X
V	Affective component of disposition, in Epstein's *Agent_Zero* framework
v	Citizens' vision radius in the ABM of social conflict
v'	Cops' vision radius in the ABM of social conflict
X	$= G - N$
x_m	Scale parameter in Pareto distribution
γ	Parameter in the proposed form of RD in the ABM of social conflict which expresses sensitivity to value (to differentiate political from economic RD)
$\mathscr{A}_v$	Set of "active" citizens within an agent's vision radius in the ABM of social conflict

$\mathscr{E}_v$	Set of empty cells within an agent's vision radius in the ABM of social conflict
P_a	Estimated arrest probability
ABM	Agent-Based model(s)
AS	"Arab Spring"
CAST	Conflict Assessment System Tool
EFA	Exploratory factor analysis
FSI	Fragile States Index (indicator published by The Fund For Peace)
FWI	Freedom in the World Indicator (published by Freedom House)
ICT	Information and Communication Technologies
ODD	'Overview, Design Concepts and Details' protocol for describing ABM
RD	Relative deprivation
RMSEA	Root Mean Square Error of Approximation
SCAD	Social Conflict Analysis Database
SN	Social networks
SOA	State-of-the-art
TLI	Tucker-Lewis Index of factoring reliability

Chapter 1
Introduction

Social conflict and violence are two fundamental characteristics of human societies [1, 61]. Both are deeply rooted on biology and evolution, and depend on cultural and situational factors [46]. As such, they are extremely heterogeneous and varied in type, motif, scale, intensity, and evolution patterns.

Understanding how the social context leads to latent social conflict and how this can evolve to large protests, civil violence or unanticipated revolution, is a topic of central importance in history, political science, sociology, social psychology, and also in military and security studies. History provides striking examples of large street protests sparking revolutions and overthrowing regimes, as in the French Revolution in 1789, the October Revolution in Russia in 1917, and the Iran Revolution of 1979 [42]. Widespread access to Social Networks (SN) and Information and Communication Technologies (ICT) changed the dynamics of social conflict processes in dramatic ways, as they allow organizations or groups of activists to summon, coordinate and show protest events or violence to a worldwide audience in almost real time [13, 25, 60].

Social conflict encompasses an extremely wide spectrum of processes and phenomena and is a multidisciplinary topic [2]. It can be latent or manifest, and manifest conflict can be non-violent (e.g., peaceful demonstrations, strikes, public petitions) or violent (demonstrations with violent confrontation, riots, and civil violence). There is no single criterion for classifying social conflict processes and phenomena. Figure 1.1 shows a simplified classification of some typical manifestations of social conflict, using the intensity of violence as a criterion and showing the scientific disciplines in which they are studied. In this figure, conflict manifestations were divided into three broad categories: (*i*) peaceful manifestations, (*ii*) low-intensity violence (non-armed conflict), and (*iii*) high-intensity violence (armed conflict).[1] This division is consistent with general definitions of "violence"

[1]"Armed conflict" is defined as *"a contested incompatibility that concerns government and/or territory where the use of armed force between two parties, of which at least one is the government*

C.M. Lemos, *Agent-Based Modeling of Social Conflict*, SpringerBriefs in Complexity,
https://doi.org/10.1007/978-3-319-67050-8_1

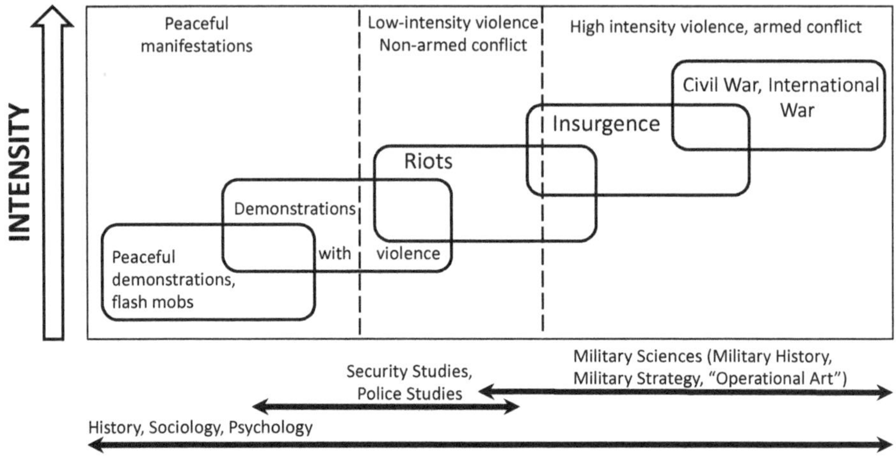

Fig. 1.1 Classification of social conflict manifestations based on their intensity (level of violence), showing the disciplines in which they are studied. Transitions between different manifestations are represented by rounded rectangles (Source: author)

[81], "civil violence" [37] and "armed conflict" [74], and also with Ted Gurr's frustration-aggression theory on civil violence [37]. There is also the possibility of escalation, represented by overlapping areas in rounded-corner rectangles shown in Fig. 1.1. Predicting the time evolution of social conflict and whether or not violence and escalation will occur, and if possible controlling these processes, is a problem of great practical importance in security studies and political science. However, these processes are very complex and path-dependent on social factors, particular settings, types of interactions and scales (of time, space, and size or proportion of individuals involved). Such processes may be stable or unstable, as sketched in Fig. 1.2. Unstable processes involve build-up of tension and can lead to sudden changes (e.g., revolution or outbreak of armed conflict) [42, 43].

The development of computers and software tools added simulation to the traditional methods of scientific research—observation, laboratory experimentation, mathematical deduction, and empirical inference [32]. Simulation has the advantages of providing a deep level of insight and understanding and allows the consideration of many different scenarios which would be impossible to test in reality. Social simulation using Agent-Based models (ABM) has been successfully introduced to study segregation [64, 65], the dissemination of culture [4], opinion dynamics [82], rumor spreading [32], and many other social phenomena [22]. Several ABM were proposed to study conflict phenomena such as civil or ethnic

of a state, results in at least 25 battle-related deaths." in page 1 of the UCDP/PRIO Armed Conflict Dataset Codebook [74]. Specific forms of armed conflict are thoroughly studied in the realm of Military Sciences. Modern definitions of "insurgence" and "war" can be found in military doctrine publications, such as the AAP6—NATO GLOSSARY OF TERMS AND DEFINITIONS [54].

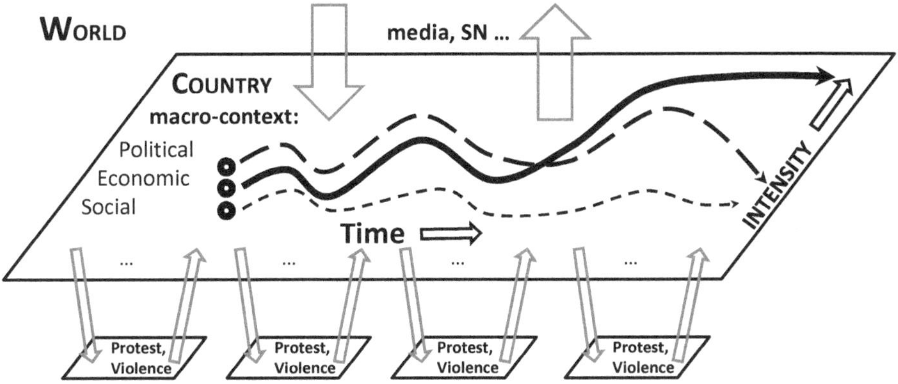

Fig. 1.2 Evolution of social conflict viewed as complex and path-dependent process with micro-macro and feedback links (Source: author)

violence [21, 24], urban crime [26], worker protest [41], riots [18], insurgence [20], and land combat [40]. ABM are suitable for studying the dynamic of protests, escalation, and civil violence, because these are large-scale, weakly organized peaceful or low-violence conflict manifestations, whereas armed conflict requires a high degree of leadership and organization [37]. Thus, the problem of modeling the dynamics of protests and violence is one of complexity and Agent-Based modeling is a promising approach for providing insight and understanding of their underlying mechanisms.[2]

In this work, an ABM for studying large-scale conflict against a central authority was developed and explored in a set of computer experiments. The model proposed herein is an extension of Epstein's ABM of civil violence, in which new mechanisms such as deprivation-dependent hardship, vanishing of the risk perception ("massive fear loss") below a critical ratio between deterrence and "group support," legitimacy feedback (drop of the central authority's legitimacy due to uprisings), and network influences were implemented.

1.1 Purpose and Scope

The purpose of the present work is to study how the complex patterns of size, duration, and recurrence of bursts of social unrest in large-scale conflict against a central authority, like those affected by the "Arab Spring" (AS), can be explained in terms of underlying mechanisms proposed by theories, and modeled via ABM.

[2]Although in the modern conceptions insurgence and war are also studied using methods from complexity sciences [40], in this work we will only consider protests and civil violence, as defined in Sects. 1.1 and 2.2, respectively.

The scope is framed by the peaceful and low-intensity conflict phenomena shown in Fig. 1.1, which involve a significant proportion of a population, and for which self-organization is important. To make this scope more precise, it is convenient to introduce the following general definitions:

Definition 1 Social conflict—Confrontation or dynamic balance of social powers, which can be latent or manifest, direct or indirect, coercive or non-coercive. (Source: author, based on Coser [14], Rummel [61] and Allan [1])

Definition 2 Demonstration—Distinct, continuous, and largely peaceful action directed toward members of a distinct "other" group or government authorities. (Source: Salehyan and Hendrix [63])

Definition 3 Riot—Distinct, continuous, and violent action directed toward members of a distinct "other" group or government authorities. The participants intend to cause physical injury and/or property damage. (Source: Salehyan and Hendrix [63])

Definition 4 Mechanism—An intelligible description and general explanation of the relationships between causes and effects in a system, in terms of the system's entities and their activities. (Source: author, adapted from Machamer et al. [47])

In this work we are only interested in politically motivated large-scale conflict against a government or authority. Other important forms of violence such as ethnic conflict (which involve identity issues [27, 68]), hooliganism, or riots mobilized by groups via Facebook, will not be treated. These forms of collective violence will be mentioned only in those aspects that are relevant for the present work. Likewise, the study and modeling of revolution will not be explicitly considered.[3]

1.2 Objectives and Research Questions

The main objective is to show that by combining Epstein's classic ABM of civil violence, with additional mechanisms postulated in Ted Gurr's frustration aggression theory and statistical analysis of datasets of conflict events, it is possible to:

- Develop an ABM with increased "generative capacity" (i.e., a richer set of meaningful behaviors) ;
- Identify key mechanisms and associated parameters (with tipping points);
- Discuss the plausibility of the solutions obtained in simulations using quantitative information about real events;

[3]Although revolution is an extremely important topic in the study of large-scale conflict (e.g., [42, 43, 48, 52]), the present work will be centered on the stability, instability, and complexity of the patterns of confrontation of a grieved population demanding change from a government, and not on the process of overthrowing governments.

and also understanding the limitations of ABM and social simulation in general. Consistent with this objective the main research questions are:

1. Which mechanisms and social context variables are the most relevant to describe the potential for social conflict and the likelihood and magnitude of large-scale protest (peaceful or violent)?
2. How can representations of these mechanisms be implemented in existing ABM of large-scale social conflict (i.e., Epstein's model)?
3. Which combinations of input parameters, distributions of agents' attributes, and functional forms of the agents' action rules lead to complex solutions (i.e., with intermittent peaks of rebellion, complicated trajectories in a suitable phase space, tipping points and changes of qualitative behavior, etc.), and how can the complexity of the solutions be associated with mechanisms?

1.3 Methodology of Development and Structure of the Work

The methodology of development consisted of three steps. First, a review of the state-of-the-art (SOA) on theories, concepts, and models of large-scale social conflict was performed. The emphasis was on the key variables and mechanisms identified in theories, particularly in Ted Gurr's frustration-aggression theory [37, 69], and in the analysis and discussion of Epstein's ABM of civil violence [21, 24]. This was followed by a statistical analysis of conflict events in eight African countries affected by the AS (Algeria, Egypt, Libya, Mali, Mauritania, Morocco, Sudan, and Tunisia), and a brief survey of some international indicators related to the potential for conflict (namely legitimacy and inequality). Finally, the ABM proposed herein was implemented and explored in a set of computer experiments, to illustrate the model's generative capability and how the newly introduced mechanisms lead to increased complexity of the solutions. Figure 1.3 illustrates the correspondence between these steps and the structure of this book, the remainder of which is structured in six chapters.

Chapter 2 contains a review of the theories and concepts relative to social conflict against a central authority. Chapter 3 contains a review of the existing ABM for simulation of large-scale social conflict. Epstein's landmark model of civil violence (Model I, [21, 24]) is described and discussed. Other ABM inspired in Epstein's ABM are also analyzed.

Chapter 4 describes an analysis of conflict events and international indicators for eight African countries affected by the AS (Algeria, Egypt, Libya, Mali, Mauritania, Morocco, Sudan, and Tunisia) based on the Social Conflict Analysis Database (SCAD) [71], the Fragile States Index (FSI) indicators by the Fund For Peace [70], the Freedom in the World Indicator (FWI) by Freedom House [28], and the "All the Ginis" database of welfare inequality [49].

Chapter 5 contains the description of the proposed "abstract" ABM for simulation of large-scale conflict against a central authority, using the "Overview, Design Concepts and Details" (ODD) protocol [36]. This is followed by a set of model

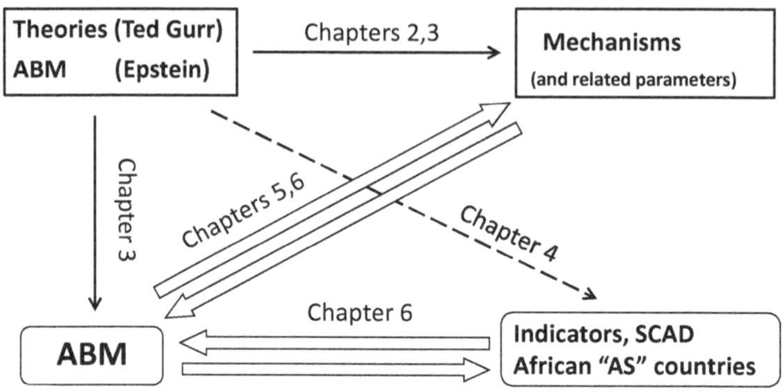

Fig. 1.3 Schematic representation of the correspondence between the steps of the methodology of development and the chapters of the present work (Source: author)

explorations, described in Chap. 6. These latter started with an analytical study of the "citizens" decision rule in Epstein's ABM to show why some forms of the estimated arrest probability function lead to solutions with large outbursts of social unrest whereas others do not, followed by a set of simulation experiments to illustrate how the newly introduced mechanisms lead to plausible results (by comparison with those obtained in Chap. 4) and influence the solutions' behavior (permanent unrest, complex solutions with intermittent peaks of rebellion, or permanent calm).

Finally, Chap. 7 summarizes the innovations and contributions of the present work to the SOA on ABM simulation of social conflict processes, of which the most salient were: (*i*) the introduction of the mechanism of "massive fear loss" (or generalized vanishing of the risk perception in the population) controlled by a parameter ρ_c that determines the size of simulated bursts of social unrest; (*ii*) the implementation of a dependency between the "hardship" and the "relative deprivation" with a parameter γ that regulates "sensitivity to value" (to model both political and economic deprivation); (*iii*) the implementation of endogenous legitimacy feedback; and (*iv*) the quantitative analysis of conflict events in African countries affected by the AS, to better understand the correspondence between reality and simulations. It was shown that both ρ_c and γ have associated tipping points. When combined with strong deprivation, legitimacy feedback does not change the qualitative behavior of the solutions but leads to interesting and complicated trajectories in the system's phase space(s). Nevertheless, legitimacy feedback induces instability and leads to solutions with alternating regime (long periods of calm followed by episodes of turmoil) in an otherwise stable scenario of high legitimacy and low level of repression. Network influences lead to instability by greatly amplifying the magnitude and increasing the duration of the simulated conflict events in otherwise stable conditions. Also, the degree of connectivity has a larger impact than the influence weight on the magnitude of the simulated conflict events.

Chapter 2
Theoretical Foundations

In this chapter a summary review of the SOA on theories of large-scale social conflict against a central authority is presented. The literature on social conflict theories is extremely vast, but the review presented herein will be focused on two theories: Ted Gurr's frustration-aggression theory on the psychological factors of civil violence [37, 38], and Gene Sharp's theory of non-violent action [66]. The first is very systematic and identifies the key variables and mechanisms of positive and negative drive for large-scale protest and violence. The second is also important because of its connection with the AS, which will be analyzed in Chap. 4. The chapter also includes a section with some notes on the concept of legitimacy, and ends with concluding remarks on the implementation of social conflict theories in ABM.

2.1 Overview of Social Conflict Theories

In general terms, conflict can be defined as a confrontation or dynamic balance of powers. Power can also be defined in very general terms as the capability of imposing will or produce a desired effect. Thus, social conflict is a confrontation of social powers (Definition 1, page 4).

Theories of social conflict played a central role in the development of sociology [1]. Karl Marx presented the first theory of social conflict based on the struggle between strata of the society (the "classes") over economic resources and the conditions that lead to such stratification. Max Weber considered the limitations of the Marxian concept of "class" (which he thought to be more complex than Marx assumed), stressed the importance of state and economy (inequality, status, party and power) for setting up conditions for social conflict, and introduced legitimacy, authority and legitimation of institutions as notions of fundamental importance for analyzing social conflict.

C.M. Lemos, *Agent-Based Modeling of Social Conflict*, SpringerBriefs in Complexity,
https://doi.org/10.1007/978-3-319-67050-8_2

Later theorists such as Georg Simmel, Lewis Coser and Ralph Dahrendorf combined elements from Marx and Weber and brought new perspectives to social conflict theory [1]. Simmel and Coser considered social conflict as an essential element in the evolution of society and the persistence of social order (whereas for previous theorists it was understood as a factor of instability and disintegration).

Coser [14] recognized conflict as an instinctual characteristic of human life (individual and social), and theorized about its sources, level of violence, and functional consequences. For this author, uneven stratification of "class," status and power, causes a sense of deprivation that leads to arousal of "class consciousness." He stated that conflicts resulting from rational goals tend to be less violent (e.g., strikes and passive resistance movements) than those about "transcendent goals" (moral or group values) and emotional involvement can be violent.

Dahrendorf [15, 16] stated that the level of violence in social conflict depends on technical, political, and social conditions of organization, as well as on effective regulation of conflict within the society. This author viewed power, not "class," as the main feature of social conflict, although he was in agreement with the ideas of dialectical change and bi-polarized society (from Marx), power and authority (from Max Weber). Dahrendorf considered that the level of violence in social conflict can be measured by the kinds of weapons used (peaceful demonstrations have a very low level of violence, whereas protests with exchange of stoning and tear gas are much more violent).

Modern theories replaced "absolute deprivation" by "relative deprivation" (RD), as the key concept for evaluating the potential for conflict. Gurr [38] defined RD as the tension due to the gap between value expectations and value capabilities and considered three patterns of RD, namely decremental, aspirational, and progressive (Fig. 2.1).[1] Progressive RD generalizes the "J-curve hypothesis" of Davies [17]—*"revolutions are most likely to occur when a prolonged period of objective economic and social development is followed by a short period of sharp reversal"*—by including expectations. Frustration associated with the potential for revolt due to

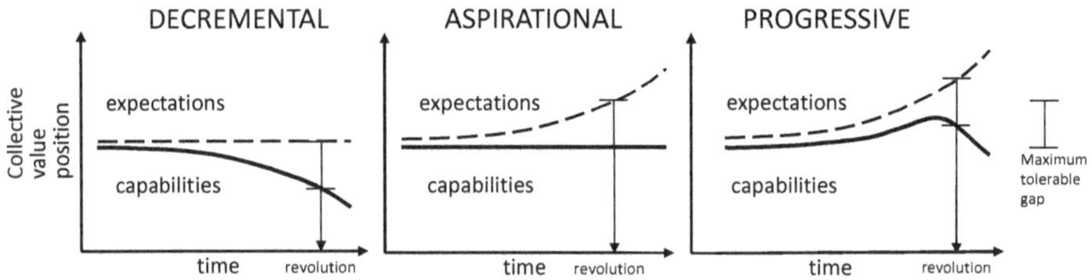

Fig. 2.1 Patterns of RD: decremental (left), aspirational (middle), and progressive (right) (Source: author, adapted from Gurr [38])

[1]Time and awareness of past own capabilities, as well as of capabilities of other groups, are also essential elements of RD.

RD leads to revolt if the level of RD increases above a tolerance threshold for a large portion of the society [9]. Bischof [9] considers three types of RD: political (e.g., deprivation of political participation in authoritarian regimes), economic (e.g., periods of economic growth followed by economic and financial crises), and social (e.g., ethnic segregation or exclusion). This author states that although economic RD is the most commonly considered type, political and social deprivation may be more important. The evaluation of RD from datasets of social indicators and variables is a very important but difficult issue.

The social conflict theories of Marx, Weber, Simmel, Coser, and Dahrendorf are based on a macro-level perspective. Randall Collins [10] introduced a micro-level perspective in social conflict theory. He also considered power and unequal distribution of resources as key features of the problem, but stressed the fundamental role of symbolic goods, emotion and ritual. In this way, he approaches the evolutionary concepts of Lorenz [46] (on aggression in animal societies) and Coser [14]. Another important point of this theory is the role of the State as holder of the legitimate use of power (and force) and the consideration of legitimacy as a special kind of emotion. Collins also analyzed conflicts in the context of history and geopolitics and concluded that:

- Unequal distribution of scarce resources (power, economic, cultural) causes potential conflict;
- Potential conflict turns to actual conflict depending on the degree of mobilization (emotional, moral, symbolic, and derived collective rituals) of the opposing groups;
- Conflict generates subsequent conflict;
- Conflict intensity lowers as mobilization resources are used.

We conclude this brief survey of social conflict theories with a reference to the conceptual approach of Rummel [61]. This author considers the following elements of social conflict: (*i*) space; (*ii*) structure; (*iii*) situation, and (*iv*) behavior. The conflict space is the multidimensional biopsychological (needs, drives, instincts) and sociocultural (religion, ethics, law, etc.) space of resources. The structure of social conflict is the sociocultural distances between individuals and groups, which generate cleavages in the society. Awareness of these distances creates disposition towards action. The conflict situation is the activation of opposing attitudes, due to expectations, arousal of needs, conflicting interests, capabilities and will of the parties. The tension created by the conflict situation may lead to manifest action or peaceful resolution. Rummel highlights the importance of triggering events and states that much of social conflict is latent and not manifest.

The conceptual approach of Rummel can be viewed as a cycle of dynamic change of power and subsequent adjustments (Fig. 2.2).

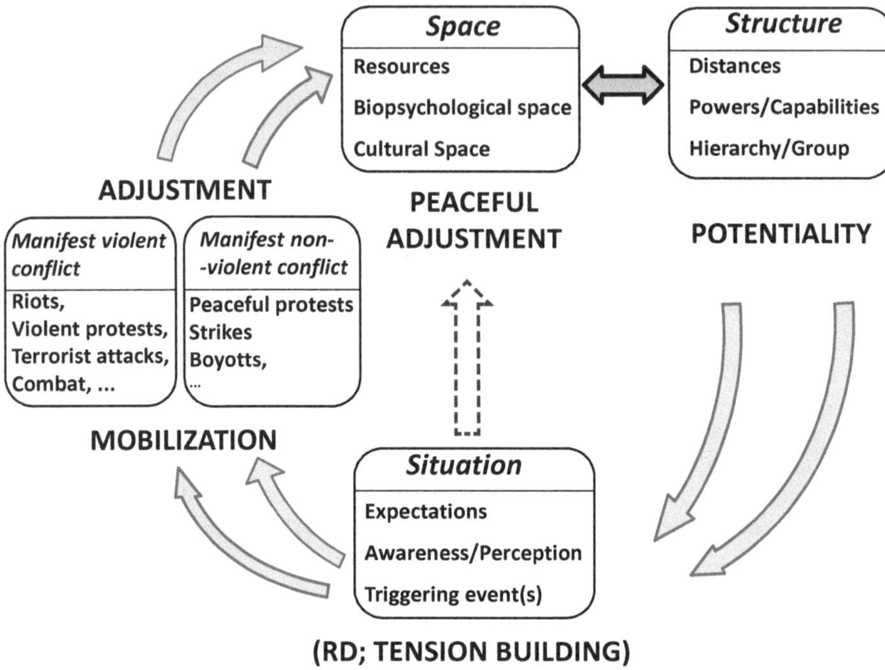

Fig. 2.2 Social conflict cycle, derived from Rummel's conceptual approach [61] (Source: author)

2.2 Violence

The study of violence is an extremely vast and difficult topic. There are two conceptions of violence [57]: (*i*) minimalistic, based on coercive use of physical force, response and harm; and (*ii*) comprehensive, which includes avoidable factors that impede human realization and violate rights or integrity. The following definition is an example of the latter type:

Definition 5 Violence—The intentional use of physical force or power, threatened or actual, against oneself, another person, or against a group or community, that either results in or has a high likelihood of resulting in injury, death, psychological harm, maldevelopment or deprivation. (Source: World Report on Violence and Health: Summary [81], page 4)

This definition includes a very wide range of acts, from self-inflicted violence (e.g., suicide) to armed conflict. In this work, we are only interested in the very general aspects of politically motivated, spontaneous and large-scale collective violence, such as occurs in demonstrations with escalation and riots (Fig. 1.1), and thus the following definition of civil violence was adopted:

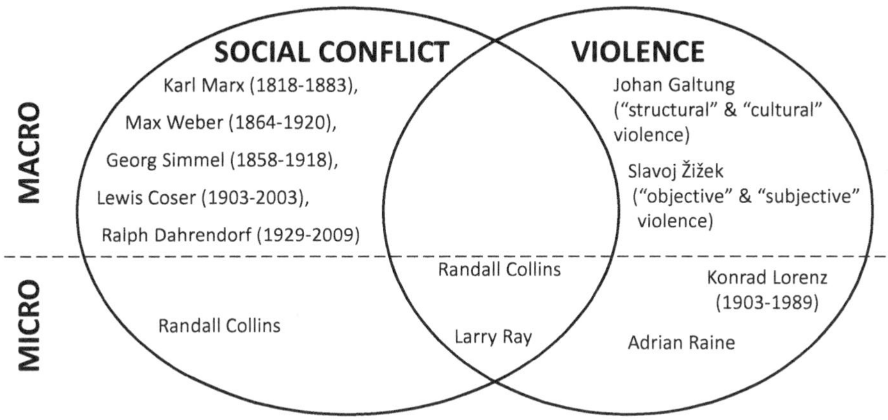

Fig. 2.3 Social conflict vs violence perspectives (Source: author)

Definition 6 Civil violence—All collective, non-governmental attacks on persons or property, resulting in intentional damage to them, that occur within the boundaries of an autonomous or colonial political unit. (Ted Robert Gurr [37], page 247)

There is a strong relationship between social conflict and violence [57], so that many theorists of social conflict (e.g., Marx, Weber, Dahrendorf, Gurr, Collins, Žizek) also presented theories on violence. Theoretical explanations for the causes of violence have been proposed from the viewpoints of evolutionary biology [46], sociology [57], psychology [7, 37], and criminology [56]. These can be classified into two broad categories: macro-oriented and micro-oriented. Figure 2.3 shows some representative authors of theories of social conflict and violence in both categories.

Macro-oriented theories are focused on structural conditions leading to collective violence, from the standpoints of sociology, economy, and historical analysis. One example is the Marxist theory of struggle between classes, according to which class division and conflict of interests of the different classes over economic resources is the main cause of conflict throughout History [57]. More modern theories are wider in scope and reflect important factors other than economical to explain violence due to social conflict. For instance, Galtung introduced the concepts of "structural violence" [29] and "cultural violence" [30]. The former is a designation for all forms by which social structures or institutions harm individuals by preventing their attainment of basic goals. The latter encompasses the aspects of a culture that can be used to legitimize direct or structural violence.[2] Žižek introduced the concepts of "subjective" and "objective" violence to distinguish between forms of violence that are "seen" and acknowledged (e.g., through "media" reports), like crime, terror, riots

[2]Nationalism, racism, and sexism are typical examples of "structural violence." Religion and ideology are potential sources of "cultural violence."

and war, and forms of violence of symbolic and systemic origin that are "unseen," i.e. which are always present, but tend to pass unnoticed in the citizens' everyday life [83].[3]

Micro-oriented theories on the causes of violence are mainly based on evolutionary biology and psychology. Gurr [37] considers three types psychological theories of aggression, namely instinctual-only, learned-only, and frustration-aggression. Instinctual theories of Freud and Lorenz are based on the assumption that aggression is a "hard wired" trait of human behavior. According to Lorenz [46] aggression has a fundamental evolutionary value for the preservation of species by ensuring the spreading over the available habitat, thus avoiding depletion of resources. Aggressive behavior results from a tension between conflicting drives, aggression (attack) and fear (escape). There is a stimulus threshold for eliciting attack which can be lowered by situational factors. Lorenz also attempts to explain crowd behavior in terms of the mechanism of "mass enthusiasm." Learned-only theories were developed mainly by child and social psychologists [37]. According to these theories, aggressive behaviors are learned and used rationally in pursuing specific goals.

2.3 Ted Gurr's Frustration-Aggression Theory on the Psychological Factors of Civil Violence

Ted Gurr's frustration-aggression theory of civil violence [37, 38] is more systematic and complete and has stronger empirical support than the ones mentioned above. It is based on the frustration-aggression hypothesis of Dollard et al. [19], according to which aggression is caused by frustration due to interference with goal-directed behavior and expectations [7, 37], and on the assumption that frustration is a function of (illegitimate) RD. Table 2.1 lists the propositions of the theory, which are divided into three groups [37]. The propositions in the first group (I) describe the potential for civil violence in terms of RD, commitment to a goal, and legitimacy. The second group (M) includes propositions relating the collective response to fear/deterrence (retribution), group protection/support, persistence of anger and memory effects, and institutionalized forms of protest (e.g., authoritarian vs democratic regimes). The last group (F) includes additional propositions on the type and scale of civil violence. Figure 2.4, adapted from [37], shows the relationship between the variables determining the potential for conflict and the possible outcomes (resignation, non-violent protest, and civil violence).

In the context of the present work, this theory is important because it links psychological and individual factors to structural factors considered in social conflict theories, identifies important variables and mechanisms that determine the

[3]Žižek's "objective" violence thus includes "structural" and "cultural" forms of violence. Unemployment, job insecurity, and cuts in public institutions that provide social welfare fit into the categories of "structural" and "objective" violence.

Table 2.1 Ted Gurr's propositions on the influence of psychological factors in the likelihood and magnitude of civil violence (Source: Ted Gurr [37])

I.1	The occurrence of civil violence presupposes the likelihood of relative deprivation among substantial numbers of individuals in a society; concomitantly, the more severe is relative deprivation, the greater the likelihood and intensity of civil violence
I.2	The strength of anger tends to vary directly with the intensity of commitment to the goal or condition with regard to which deprivation is suffered or anticipated
I.2a	The strength of anger tends to vary directly with the degree of effort previously invested in the attainment or maintenance of the goal or condition
I.2b	The intensity of commitment to a goal or condition tends to vary inversely with its perceived closeness
I.3	The strength of anger tends to vary inversely with the extent to which deprivation is held to be legitimate
I.4	The strength of anger tends to vary as a power function of the perceived distance between the value position sought and the attainable or residual value position
I.5	The strength of anger tends to vary directly with the proportion of all available opportunities for value attainment with which interference is experienced or anticipated
M.1	The likelihood and magnitude of civil violence tend to vary curvilinearly with the amount of physical or social retribution anticipated as a consequence of participation in it, with likelihood and magnitude greatest at medium levels of retribution
M.1a	Any decrease in the perceived likelihood of retribution tends to increase the likelihood and magnitude of civil violence
M.2	Inhibition of civil violence by fear of external retribution tends in the short run to increase the strength of anger but in the long run to reduce it
M.2a	The duration of increased anger under conditions of inhibition tends to vary with the intensity of commitment to the value with respect to which deprivation is suffered
M.3	The likelihood and magnitude of civil violence tend to vary inversely with the availability of institutional mechanisms that permit the expression of non-violent hostility
M.4	The likelihood and magnitude of civil violence tend to vary directly with the availability of common experiences and beliefs that sanction violent responses to anger
M.4a	Given the availability of alternative experiences and beliefs, the likelihood that the more aggressive of them will prevail tends to vary with the strength of anger
M.5	The likelihood and magnitude of civil violence tend to vary directly with the extent to which the deprived occupy organizational and/or ecological settings that provide (1) normative support through high levels of interaction, (2) apparent protection from retribution, and (3) congruent models for violent behavior
F.1	The characteristic form of civil violence tends to vary with the differential incidence of relative deprivation among elite aspirants and masses: (1) mass deprivation alone tends to be manifested in large-scale civil violence with minimal organization and low intensity; (2) elite-aspirant deprivation tends to be manifested in highly organized civil violence of high intensity
F.1a	Whether organized and intense civil violence is large-scale or small scale is a joint function of the extent of mass deprivation and strategic access of deprived elite aspirants to the incumbent potential elite

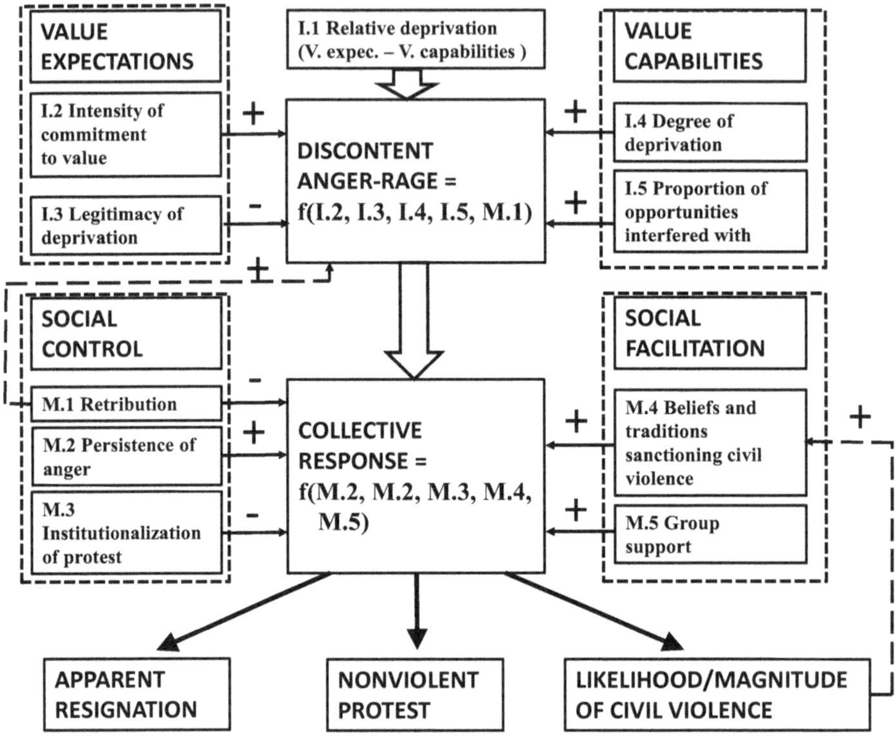

Fig. 2.4 Variables determining the likelihood and magnitude of civil violence (Source: Ted Gurr [37])

likelihood, magnitude and type of civil violence, and is very systematic. Inspection of Table 2.1 and Fig. 2.4 shows that:

- The three sets of propositions in Table 2.1 and the upper, middle, and bottom blocks in Fig. 2.4 bear a correspondence with the potential/mobilization/outcome sequence of the cycle sketched in Fig. 2.2;
- RD, commitment, and legitimacy are important factors of aggression potential;
- Deterrence (perceived likelihood of retribution) and institutional mechanisms for expression of non-violent confrontation inhibit collective violence;
- Group protection and "energizing" enhance collective violence;
- Massive RD within a population tends to result in large-scale, "minimal organization" and low-intensity violence, whereas deprivation among "elite aspirants" tends to result in organized, high-intensity violence.

This last proposition conveys some theoretical support for the classification of conflict phenomena sketched in Fig. 1.1. It is also consistent with the idea that the methods of complex systems studies are suitable for studying conflict phenomena at the low end of the violence spectrum.

2.4 Gene Sharp's Theory of Non-Violent Action

In his extremely influential book "From Dictatorship to Democracy, A Conceptual Framework for Liberation" [66], Gene Sharp considered the problem of overthrowing authoritarian regimes without allowing new dictatorships to emerge, and developed his well-known theory of non-violent action. Thus, according to Gene Sharp, resistance via non-cooperation and non-violent action is the only effective way of fighting dictatorships.

This theory has many points of interest for the present work. First, non-violent action is a complex process, involving a diversity of methods (in Annex 1 of [66], Sharp lists 198 different methods of non-violent action, particularly demonstrations, strikes, and boycotts). Second, large and continued protests tend to generate a collective sense of fear control in the population, which can be interpreted as a hypothetical mechanism of "massive fear loss" once protests become sufficiently large and frequent. Furthermore, large and continued protests weaken the government's strength, but may lead to brutal repression and further loss of legitimacy. Third, the existence of strong, independent social groups and institutions, and careful strategic planning are vital for the success of non-violent fight against dictatorships. Thus, Sharp's theory of the fight for democracy in authoritarian regimes blends considerations of complexity with hierarchical and strategic thinking.

2.5 Legitimacy

Legitimacy is the value which confers recognition and acceptance of the exercise of power by an authority or institution. It plays a key role in the theories on social conflict and civil violence, for governments perceived as illegitimate by a population are unstable and usually overthrown.

There is no generally accepted formula for evaluating the legitimacy of states. Beetham [6] considers that legitimacy can be expressed in terms of three main components: legality, normative justifiability, and expressed consent.

Following this three-component conception, Gilley [33] devised a formula for computing an aggregate legitimacy score for 72 countries in terms of sub-indicators related to three subtypes of legitimacy: *views of legality* (related to formality and law), *views of justification* (related to moral beliefs), and *acts of consent* (related to behavior). This author argues that justification has been the most underestimated legitimacy subtype and proposes a weighted average legitimacy score with 50% weight for *views of justification* and 25% for *views of legality* and *acts of consent* [34].

2.6 Concluding Remarks

The main conclusions of the above review of theoretical foundations of social conflict against a central authority can be summarized as follows:

- The two key concepts that determine the potential for conflict are RD and legitimacy. RD is the primary source of conflict potential, and can be political, social, or economic;
- Legitimacy is variable that confers acceptance of the exercise of power to an authority. As such, it plays a central role, because RD must be perceived as illegitimate to be a source of conflict;
- Two theories were reviewed, Ted Gurr's RD-based frustration-aggression theory on the psychological factors of civil violence, and Gene Sharp's theory of non-violent action. The first is very systematic in its identification of the important variables and mechanisms. The second has been considered as an inspiration for large-scale conflict processes such as the AS, and contains premises on the role of self-organization and a mechanism of "massive fear loss" due to large and continued protests;
- Gilley's theoretical framework on the meaning and measure of legitimacy expresses this latent concept as a weighted average of three components: namely views of legality, views of justification, and acts of consent. This approach provides a starting point for implementation of legitimacy feedback in ABM.

Chapter 3
Review of Agent-Based Models of Social Conflict and Civil Violence

This chapter contains a review of the SOA on ABM for simulation of large-scale social conflict and violence. Existing ABM are focused on specific scales and types of conflict phenomena, and according to Gilbert [31] can be classified as "abstract," "middle-range," and "facsimile" models. "Abstract" ABM describe emergent properties of complex systems in terms of a small number of agent types, simple rules, and interactions. "Middle-range" ABM are used to describe phenomena with definite space and time scales. The purpose of "facsimile" ABM is the description of the system's dynamics with as much realism as possible. The present review will be centered on "abstract" ABM, which are of primary interest given the scope and purpose of the present work (Sect. 1.1).

The present chapter is organized in four parts. The first contains a summary of definitions and concepts related to agents, agent types, and rule-based models of binary decision (e.g., rebel or not; adopt an innovation or not; attack or remain passive; etc.) used in ABM of "abstract" type. The second contains the presentation and discussion of Epstein's ABM, which is a landmark model of civil and ethnic violence due to its simplicity, soundness, and explanatory power. The third part contains a review of extensions of Epstein's ABM that have been proposed by several authors. The chapter ends with some remarks on the limitations of existing ABM and on possibilities for their improvement.

3.1 Agent Definition, Agent Types and Rule-Based Models of Binary Decision

In this work, the following definition of "agent" was adopted:

Definition 7 Agent—An *agent* $\mathscr{A}$ is a computer system that is situated in some environment and is capable of perceiving, deciding, and performing actions in an autonomous way. In more abstract terms, if $E = \{e, e', \ldots\}$ is the set of possible

C.M. Lemos, *Agent-Based Modeling of Social Conflict*, SpringerBriefs in Complexity, https://doi.org/10.1007/978-3-319-67050-8_3

environment states; $Ac = \{\alpha, \alpha', \ldots\}$ is the set of actions available to $\mathscr{A}$; $r = (e_0, \alpha_0, e_1, \alpha_1, \ldots)$ is a *run* (sequence of environment states alternating with actions by $\mathscr{A}$); $\mathscr{R}$ is the set of possible finite runs; $\mathscr{R}^{Ac}$ is the subset of $\mathscr{R}$ ending with an action; and $\mathscr{R}^{E}$ is the subset of $\mathscr{R}$ ending with an environment state; then an *agent* is a function that maps runs ending in environment states to actions [80]: $\mathscr{A} : \mathscr{R}^{E} \rightarrow Ac$. (Source: Wooldridge [80])

Russel and Norvig [62] consider four agent types: (*i*) simple reflex (reactive), which respond immediately to percepts; (*ii*) agents with internal state, which respond based on percepts and on tracking of their internal state; (*iii*) goal-based agents, which try to achieve goals (by means of searching and planning); and (*iv*) utility-based agents, which try to maximize an utility function. "Abstract" ABM are mainly based on the first two types, since agents only perform very simple actions and do not need to be endowed with sophisticated reasoning capability (for planning and deciding).

Threshold rules have been widely used to model problems of binary decision ($Ac \in \{0, 1\}$, where 0 means "not act" and 1 "act") involving a cost-benefit balance that depends on individual factors and external influences. These rules have been used in "abstract" ABM of complex systems based on reactive agents to simulate social phenomena such as sorting in segregation [64], behavioral contagion in rioting [35], peer pressure (or majority influence towards conformity [3, 27]), standing ovations (SOV) [50], and civil violence [24].

Epstein proposed an agent model (*Agent_Zero*) of binary action based on a threshold rule, which is an additive combination of three components, affective (V), deliberative (P), and social (S) [23]. The i-th agent *solo disposition* D_i^{solo}, *total disposition* D_i^{total} and action rule are given by the following equations [23]:

$$D_i^{\text{solo}}(t) = V_i(t) + P_i(t) \tag{3.1}$$

$$D_i^{\text{total}}(t) = D_i^{\text{solo}}(t) + \sum_{j \neq i} \omega_{j,i} D_j^{\text{solo}}(t) \tag{3.2}$$

$$\text{Action Rule : Act iff } D_i^{\text{tot}} > T_i \tag{3.3}$$

where T_i is the threshold of the i-th agent and the weights $\omega_{j,i}$ represent a dispositional influence network [23]. These equations model the mechanism of *dispositional contagion* instead of *behavioral imitation*, since the state of other agents does not appear as a variable in equations (3.1)–(3.3). Disposition contagion successfully represents the phenomenon of an agent that initiates action without imitating any other agent, and according to Ted Gurr is more important than imitation for explaining civil violence:

> *Frustration-aggression theory is more substantially developed, however, and has substantially more empirical support than theories that assume either that all men have free-flowing source of destructive energy or that all aggression is imitative and instrumental.* - in *Why Men Rebel* [38], page 33.

The affective component of the i-th agent is modeled according to a generalized Rescorla-Wagner theory of conditioning [58]:

$$\frac{dV_i}{dt} = \alpha_i \beta_i V_i^{\delta}(\lambda - V_i) \tag{3.4}$$

where V_i is the associative value of the conditioned stimulus CS (e.g., attack–anger) for agent i, $\alpha_i \in [0, 1]$ and $\beta_i \in [0, 1]$ are learning constants, $\delta \in [0, 1]$ is an exponent which determines the form of the "learning trajectory",[1] λ is the maximum associative value of the CS and t is time. Equation (3.4) describes both acquisition and extinction of emotional disposition, by setting λ different or equal to zero, respectively.

3.2 Epstein's Agent-Based Model of Civil Violence

The ABM by Epstein et al. [24], herein referred to as Epstein's model [21], is a successful and popular model for simulation of civil violence. The present review of this model includes the following items: purpose, entities, process overview, and scheduling and submodels. These are a subset of the "Overview, Design Concepts and Details" (ODD) protocol [36].

Purpose The purpose of the model is the description of the dynamics of large-scale decentralized uprisings. The model has two variants of civil violence: rebellion against a central authority (Model I) and ethnic violence between two rival groups mediated by a central authority (Model II).

Model Entities The model includes two types of agents, "citizens" and "cops." The environment is a 2D homogeneous torus space. Tables 3.1 and 3.2 show the global parameters and agent attributes, respectively.

Process Overview and Scheduling In this model, "citizens" remain "quiet" or become "active" (rebellious) depending on their grievance towards the central authority (in Model I) or the rival group (in Model II), and on their risk perception. "Cops" represent the central authority and try to keep order by arresting "active" citizens. "Citizen" and "cop" agents have one move and one action rule. The move rule is the same for "citizens" and "cops":

Rule M: move to a random empty cell within the agent's vision radius.

[1] $\delta = 0$ implies an exponential variation, $\delta = 1$ a logistic (S-curve), and $\delta \in (0, 1)$ other S-curves which are not analytically solvable (see [23], page 67).

Table 3.1 Global model parameters in Epstein's ABM and their values/ranges taken from Appendix A to Epstein et al. [24] (Source: Epstein et al. [24])

Parameter	Value
Grid dimensions	40×40
Initial population density	0.7
Initial cop density	0.04
Legitimacy, L	0.8–0.9
Arrest probability constant, k	2.3
Population threshold, T	0.1
Max. Jail term, J_{max}	15–∞
Max. Age (Model II)	200
Cloning probability (Model II)	0.05

Table 3.2 Agent attributes in Epstein's ABM and their values/ranges taken from Appendix A to Epstein et al. [24] (Source: Epstein et al. [24])

Agent type	Parameter name	Variable name	Value/Range
"citizen"	Vision radius	v	1.7–7
	Hardship	H	$\sim \mathscr{U}(0, 1)$
	Risk aversion	R	$\sim \mathscr{U}(0, 1)$
	Grievance	G	$= H \times (1 - L)$
	Group (Model II)	*group*	"Blue" or "Green"
	Death age (Model II)	death_age	$\sim \mathscr{U}(0, \text{max\_age})$
	Cloning probability (Model II)	p	0.05
"cop"	Vision radius	v'	1.7–7

The (threshold) action rule for "citizen" agents describes the cost-benefit balance of turning rebellious:

Rule A: if $G - N > T$ be "active"; otherwise be "quiet" (Model I)
if $G - N > T$ kill one agent of the rival group within the vision radius; otherwise be "quiet" (Model II)

where $G = H \cdot (1 - L)$ is the grievance, $H \sim \mathscr{U}(0, 1)$ is the perceived hardship, L is the perceived legitimacy[2] assumed equal for all agents, $N = R \cdot P_a$ is the net risk perception, where $R \sim \mathscr{U}(0, 1)$ is the risk aversion, P_a is the estimated arrest probability, and T is a threshold (assumed constant for all "agents"). The form of the arrest probability presented in [24] and [21] is

$$P_a = 1 - \exp(-k \cdot (C/A)_v) \tag{3.5}$$

[2]In Model I L is the legitimacy of the central authority; in Model II L is the group's assessment of the legitimacy of the rival group to exist, assumed equal for both groups [21, 24].

where C_v and A_v are the number of "cops" and "active" agents within the agent's vision radius v and $k = 2.3$ is the arrest constant [21, 24]. Wilensky [78] proposed to replace equation (3.5) by the alternative form

$$P_a = 1 - \exp(-k \cdot \lfloor (C/A)_v \rfloor) \tag{3.6}$$

which leads to a drop of P_a from 0.9 to zero for $C_v < A_v$ and produces solutions with large intermittent bursts of rebellion [78]. Other forms for the estimated arrest probability with properties similar to those of equation (3.6) have been proposed by other authors [18, 26]. The action rule for "cops" is

Rule C: Inspect all sites within v' and arrest a random "active" citizen

Arrested citizens are removed from the simulation space ("jailed") for $J \sim \mathcal{U}(0, J_{\max})$ cycles (jail term). The jailing of citizens introduces a memory effect in the system, with a time scale proportional to $J_{\max}$. All agents are activated once per cycle in random order. Since Model II involves killing of "citizen" agents, a simple form of population dynamics is implemented in this case: "citizens" clone offspring with probability p upon activation, and die after *death_age* cycles (Table 3.2).

Main Results Epstein's ABM successfully describes many characteristics of large-scale civil violence: large intermittent peaks of rebellion (punctuated equilibrium), the effects of sudden or gradual drops of legitimacy (controlled increase of the number of jailed "citizens" in the first case and a sudden large burst of rebellion in the second case), or progressive "cop" reductions (a sudden burst of rebellion due to the central authority neglecting its deterring capability below a critical point). The strength of Epstein's model lies in its explanatory power, which derives from the simplicity of the formulation and the relevance of the variables chosen for representing the social context and individual factors (Sects. 2.1 and 2.5).

Submodels The submodels in Epstein's ABM are (*i*) the expression of grievance as the product of hardship by the perceived illegitimacy, (*ii*) the expression of the net risk as the product of risk aversion by the estimated arrest probability, and (*iii*) equation (3.5) for the estimated arrest probability.

3.3 Discussion of Epstein's Agent-Based Model

Given the importance of Epstein's ABM of civil violence, it is appropriate to discuss it in greater detail with respect to three aspects: scope, mechanisms, and scales. More specifically, the key questions for such discussion are:

- Is the model applicable to civil violence only, or to both peaceful and violent conflict phenomena (demonstrations and riots, in Fig. 1.1)?
- Which quantities should be used to compare the duration, interval and size of conflict events in real processes and model simulations?

– Which parameters or combinations of parameters are related to scales (of time, space or artificial population), and how do they influence the behavior of the solutions?

3.3.1 Scope

The scope of the model, i.e. its applicability to peaceful or violent conflict manifestations, is related to the micro-scale mechanisms of individual decision and interaction, and to the macro (system level) memory induced by jailing. The threshold decision rule of "citizens" expresses a conflict between opposing drives, the impulse to rebel (similar to what Kuran [42] calls the "private preference") associated with G and the fear of retribution expressed by N.

The impulse to manifest opposition to the central authority exists in both peaceful demonstrations and riots. If peacefully showing opposition to the government entails a perceived risk, and law-enforcing agents effectively harm or arrest citizens that peacefully challenge the government (which is the case in dictatorships), then the model can be used to study both peaceful and violent forms of self-organized large-scale conflict (mainly peaceful demonstrations and riots), although the values of the parameters G, R, and T and the arrest probability P_a may be different in each case.[3] Ted Gurr's frustration-aggression theory on the psychological factors of civil violence (Fig. 2.4) is fully consistent with this argument: the mechanisms that relate discontent to collective response are the same, but the manifest response may be either peaceful or violent.

3.3.2 Measures of Size, Duration, and Interval (Waiting Time) of Outbursts

In Epstein's ABM the time scale is indefinite. In abstract terms, one model cycle can be thought of as the time interval for all "citizen" agents in the artificial society to move and decide their next state, and for "cops" to react. In real societies this depends on how fast the information about the state of other citizens reaches and influences a particular (randomly chosen) citizen and agents of authority respond. In today's highly connected world a time scale of one day is a sensible choice for

[3]This situation is similar to, e.g., the use of the diffusion equation for modeling both heat and mass transfer in Physics. The mechanism of diffusion at the molecular scale is the same, so that it can be represented by an equation of identical form, with different parameters in the constitutive equation that relates the heat or mass flow to the gradient of temperature or concentration, respectively.

large conflict events that recur intermittently over longer periods.[4] One possible way of defining a correspondence between one model cycle and the characteristic time scale of duration and interval between conflict events is to compare dimensionless quantities, such as the proportion of total time with calm and unrest, in real processes and simulations.

The absolute size (number of participants) in large demonstrations or riots is perhaps the most important element in terms of news impact. However, the proportion or % of the population involved in those events is a better measure of the society's mobilization, and also more appropriate for comparing related processes in different countries.

3.3.3 *Input Parameters, Scales and Mechanisms*

The input parameters in Epstein's ABM are related to three contexts (or scales): individual (agent), local (cells within the agent's vision radius), and global (system). They also influence the mechanisms of emergence and recurrence of large intermittent peaks of rebellion, and consequently the size, duration, and interval between these simulated bursts of activity.

Legitimacy, Estimated Arrest Probability and Threshold The decision rule also involves L and P_a, which are global and local, respectively. Legitimacy directly affects the grievance of the whole population. The estimated arrest probability expresses the effect of the local context on the risk perception (fear of retribution). To determine whether or not large cascades of rebellion will occur, it is necessary to consider the distribution of $G - N$ and the value of T. The threshold T can be interpreted as a "barrier" of perceived risk that cannot be lowered by collective behavior. In real conflict processes, it is a function of the level of repression and collective experiences and beliefs about violent response by the government (in the spirit of proposition M.4 of Ted Gurr's frustration-aggression theory of civil violence, shown in Table 2.1, page 13).

Vision Radii (v and v') For "citizen" agents, the variance of the ratio $(C/A)_v$ of agents in equations (3.5) and (3.6) decreases for increasing vision radius v. Thus, it can be expected that small v increments the probability of occurrence of localized bursts of rebellion, triggered by agents in the population with large $G - N$. For "cop" agents, both the detection capability and the mobility increase with v'. The difference between v and v' can be interpreted as a difference between the sphere of influence of "citizens" in their random contacts, and the information space of the law-enforcing agents. Hence, v is related to the triggering (emergence) and size of the rebellion peaks, and v' (together with the number or density of "cops") to

[4] Also, one day is the time unit for the duration and interval between events (start date) in databases of conflict events [63].

their suppression (duration of rebellion peaks). Thus, it can be expected that the difference between v and v' has a significant impact on the solutions' behavior.

Deterrence In Epstein's ABM deterrence is a function of "citizen" agents' local conditions via $P_a(C/A)$. The parameters that influence deterrence capability are the threshold (considered above), the density or number of "cop" agents, v' and $J_{\max}$. The influence of progressively reducing the number of "cops" was studied in Epstein et al. [24], and Epstein [21], but the effect of the other two parameters on the solutions' behavior was not investigated in a systematic way. The probability of detecting an "active" citizen in a randomly chosen cell depends on the density (or number) of "cops" and on v'. If the union of the "cops" individual information spaces does not cover the whole simulation space (due to an insufficient number of "cops", small v', or both), then rebellion bursts can start and grow undetected. This explains the sudden rebellion burst as the number of "cops" is progressively reduced [21, 24].

3.4 Other Models Based Epstein's Agent-Based Model

Kim and Hanneman [41] proposed and ABM of worker protest with "citizens" replaced by "workers" in which the grievance is expressed as a function of RD resulting from wage inequality, and state changes (between "quiet" and "active") depend on group identity effects. This approach is well founded, since RD and group grievance are positively correlated with the potential for conflict [38, 67]. The model has one global variable, the "wage dispersal" *WD*, and "worker" agents have three more attributes with respect to Epstein's model: wage $w \sim \mathcal{N}(WD/2, (WD/6)^2)$, "tag" $t \simeq \mathcal{U}(0,1)$, and "tolerance" $T \sim \mathcal{N}(1/2, 1/6^2)$. In this model, RD is computed as the difference between the ego's wage and the average wage of visible "workers," and the grievance is expressed in terms of RD using a function that gives zero grievance for ego's wage above average and produces a sharp increase of G for small negative values of RD. For agent i, an agent j within vision radius is labeled as "us" if $|t_i - t_j| < T_i$ and "them" otherwise. If group identity is not taken into account, "workers" change state according to Rule A (Sect. 3.2, page 20). If group effects are taken into account, the number of visible "us" must be greater, then the number of visible "them" for "workers" to turn "active," in addition to Rule A. Solutions obtained with this model showed punctuated behavior as in Epstein's Model I, and protest frequency was more influenced by wage inequality than by group tag effects. This model introduces important ideas, such as RD and group effects. However, the authors did neither present a clear theoretical justification for the chosen form of $G = f(RD)$, nor attempt any empirical validation of their results.

Fonoberova et al. [26] used Epstein's model to study the dynamics of crime and violence in urban settings. The purpose of this study was to determine the dimension of a police force necessary to keep crime and violence in a city below a certain threshold level. These authors considered: (*i*) different forms of the arrest probability function with monotonic and non-monotonic variation; (*ii*) the effect of "citizens" that never change state; (*ii*) different grid sizes (to simulate the conditions in small and large cities), population densities and number of "law enforcement officers" per 1,000 inhabitants. The model simulations were compared with datasets on crime and violence in 5,560 U.S. cities.

The work by Fonoberova et al. [26] has important and distinctive aspects, such as highlighting the importance of the form of P_a, the use of large numbers of agents and real data for model validation, and systematic exploration of the 3D parameter space of "citizen" density, number of "law enforcement agents" and population size. However, it also has significant drawbacks. For instance, the grievance is expressed in terms of legitimacy, which is very questionable in a model of urban crime. The authors did not explain why non-monotonic forms of P_a significantly change the solutions' behavior. In the exploration of the parameter space, the vision radius of "citizens" was set to $v = \sqrt{N_c/(\pi * \texttt{citizens density})}$, where N_c is a fixed number of nearest neighbors of a "citizen," and $v = v'$, but the authors did not investigate the solutions' behavior as a function of these parameters.

Epstein et al. [24] suggested a mechanism for introducing endogenous legitimacy feedback in Model II (ethnic violence), but this approach was neither implemented nor further explored for the case of civil violence (Model I). Lemos et al. [44] implemented endogenous legitimacy feedback in Epstein's Model I, based on Gilley's approach described in Sect. 2.5. It was found that legitimacy feedback leads to interesting effects, such as intermittent alternations of regime (very long periods of calm followed by very long periods of unrest) or apparently stable authoritarian regimes facing an unexpected massive uprising and struggling afterwards with intermittent rebellion because they never recover their initial legitimacy [44]. Also, heterogeneous legitimacy feedback leads to more unstable solutions than homogeneous legitimacy feedback, showing that non-linear effects are important.

Recently, Alessandro Moro proposed an ABM for the dynamics of violent political revolutions with three types of agents: citizens, armed revolutionaries, and agents of a central authority [52]. Although armed uprisings are not within the scope of the present work, the model by Moro is inspired in Epstein's Model I and has interesting features. Citizens may be quiet or rebellious, armed revolutionaries try to kill agents of the central authority, and agents of the central authority try to maintain order by arresting rebellious citizens and killing armed revolutionaries. The purpose of the model was to describe the patterns of the time evolution of revolutions, including (*i*) pre-revolutionary spontaneous riots, (*ii*) rebellion, and (*iii*) its different outcomes (revolution, or overthrowing of the regime, failed rebellion or anarchy/destabilization), for different initial scenarios. It is assumed that citizens turn rebellious motivated by poor economic conditions, according to the following rule [52]

Rule C': if $G(y_i) - N(y_i) > f$ be "active"; otherwise be "quiet" where $G(y_i)$ and $N(y_i)$ are the grievance and net risk, which depend on the agent's income y_i. The income is a random variable modeled using a log-normal distribution $d(y_i) = 1/(by_i\sqrt{2\pi})exp(-(\ln(y_i)-a)/(2b^2)),\ y_i > 0$, where a and b are constants. The expressions for $G(y_i)$ and $N(y_i)$ are

$$G(y_i) =(1 - L)H(y_i) \tag{3.7}$$

$$N(y_i) =A_iJ(y_i) \tag{3.8}$$

where L is the legitimacy, $H(y_i) = \exp[E(y_i)-y_i]/(1+\exp[E(y_i)-y_i])$ is the hardship index, $A_i = 1 - \exp(-w_1P_v/(1 + C_v + R_v))$ is the estimated arrest probability, and $J(y_i) = 2\exp(w_2y_ijmax)/(1+\exp(w_2y_ijmax))-1$ is the risk aversion, where P_v, C_v, and R_v are the number of visible agents of the central authority, "active" citizens and revolutionaries, respectively, and w_1, w_2, and $jmax$ are constants. In this formulation, the hardship index models economic RD using a logistic function, the risk aversion is a function of the fear of losing income whilst at jail, and the estimated arrest probability is a straightforward extension of the expression used in Epstein's model. The modeling of the armed revolutionaries and agents of the central authority is very simple and straightforward [52].

Although the ABM presented by Moro has important features such as the representation of grievance as a function of economic RD, it also has some limitations. For instance, there is no attempt to model political RD, which was the key factor of conflict potential in many cases (e.g., the AS [9]), or to use data or indicators from real processes for parametrization or validation. Also, the values of the grid size, initial densities of citizens and agents of authority, vision radius v and maximum jail term $jmax$ used in the simulations are the same as in Epstein et al. Run 2 [24]. The legitimacy was set $L = 0.85$ in all simulations, which is hardly representative of governments that face revolutions, regardless of the measure or concept of legitimacy adopted.

3.5 Concluding Remarks

Epstein's ABM of civil violence (Model I) played a central role in generative social simulation of conflict due to its simplicity, soundness, and explanatory power. It can be applied to civil violence and peaceful manifest conflict, provided the relevant mechanisms are the same in both cases. However, it has the drawbacks of not representing RD or commitment to value (to distinguish political from economic deprivation), and the mechanisms of legitimacy feedback and network influences, all of which are important for modeling real conflict processes. Other authors improved Epstein's model, but all the models have significant limitations. This suggests the following ways for improving the existing ABM:

- Include mechanisms such as legitimacy feedback and network influences using simple extensions of the "citizens" action rule in Epstein's model;

- Explore which input parameters (and related mechanisms) have associated tipping points, and explain how the latter are consistent with the theories and patterns of conflict events inferred from datasets;
- Investigate possible forms for modeling hardship in terms of RD and compare the simulated results with real cases (e.g., time evolution of protests and riots during the AS).

Chapter 4
Analysis of Conflict Datasets and Indicators: The Case of the "Arab Spring"

This chapter contains an analysis of selected indicators of legitimacy, human rights and inequality, as well as information about conflict events for eight African countries affected by the AS, namely Algeria, Egypt, Libya, Mali, Mauritania, Morocco, Sudan, and Tunisia. The AS was a recent and very important large-scale conflict process against the central governments of several Arab countries in Africa and the Middle-East, whose main issue was the struggle for human rights and political liberties, and is therefore relevant in the context of the present work. The African countries mentioned above were selected because they were the ones for which information on both indicators and conflict events was available. They are also important due to the magnitude of the publicized events, the impact of the conflict process on the respective societies (violence, regime changes, and even war), and the quick propagation of the uprisings from country to country. The main purpose of the analyses was to obtain plausible estimates for the input parameters and patterns of size, duration and recurrence of conflict events that could be used for the parametrization, validation, and exploration of the ABM.

4.1 Analysis of the Social Conflict Analysis Database for the African "Arab Spring" Countries

The SCAD database [71] contains 42 items of information (fields) about 16730 social conflict events (protests, strikes, anti- and pro-governmental violence, etc.) for 61 countries in Africa, Central America, and the Caribbean region from 1990 to 2013 based on an extensive compilation of news from Associated Press and *Agence France Presse*, via the Lexis-Nexis news service. The database is provided as two CSV files and one Codebook which describes the information item, type (date, numeric, categorical, Boolean, etc.), and coding for each field of the database [63].

C.M. Lemos, *Agent-Based Modeling of Social Conflict*, SpringerBriefs in Complexity, https://doi.org/10.1007/978-3-319-67050-8_4

Table 4.1 Summary of the information items (fields) in the SCAD database (Source: Salehyan and Hendrix [71])

Field	Type	Description
countryname	Text	Country name
startdate	Date	Start date
duration	Numeric	Duration (days)
etype	Categorical (nominal)	Type of event
escalation	Categorical (nominal)	0 for no escalation; 1–10 otherwise (according to type of the subsequent event)
cgovtarget	Boolean	*true* (1) if Central Government was targeted, *false* (0) otherwise
npart	Categorical (ordinal)	Estimated number of participants
ndeath	Numeric	Number of death (if any)
repress	Categorical (ordinal)	No repression, non-lethal repression or lethal repression used
locnum	Categorical (nominal)	Coding of event location
issue1	Categorical (nominal)	Main issue that caused event
issuenote	Text	Brief description of event

Table 4.1 shows a summary of the information items that were used in the present work.

The information in SCAD allows the retrieval of the time evolution of conflict events by type and estimated number of participants, as well as its relation with actor(s), target(s), issue(s), escalation, geographic location (urban, rural, multiple or single, regional or nationwide, etc.) and use of repression by the government. This can be done efficiently for one country or a group of countries in a given period using functions coded in R [55] (Table 4.2).

The R functions were used to analyze the conflict events that occurred before and after the AS in Algeria, Egypt, Libya, Mali, Mauritania, Morocco, Sudan, and Tunisia.[1] The general questions for exploration were:

1. How important were demonstrations and riots, in terms of number of events and estimated number of participants?
2. Which were the issues, organization, and escalation in large demonstrations and riots?
3. What were the patterns of recurrence, duration, and size of demonstrations and riots?

and for each of these questions:

[1]The current version of SCAD does not include information about conflict events in other countries affected by the AS such as Bahrain, Jordan, Kuwait, Oman and Syria. Also, there is no information about conflict events in European countries involved in the European Sovereign Debt Crisis, which would be worthwhile studying and comparing with those in the countries affected by the AS.

Table 4.2 R functions for exploration of the SCAD database (Source: author)

R function	Usage
`get.events.in.database()`	Construct a data frame from CSV files
`get.country.list(scad.data.frame)`	Get list of countries in a SCAD data frame
`get.country.events(scad.data.frame,country.names)`	Retrieve events for one or more countries
`get.dates.range(scad.data.frame)`	Get time limits of events in a SCAD data frame
`get.events.between.dates(scad.data.frame, from.date,to.date)`	Get SCAD data frame with events between from.date and to.date in an input data frame
`get.events.by.category(scad.data.frame, key,values)`	Get a subset of a SCAD data frame
`get.categories(scad.data.frame)`	Get a list of user-defined categories (used for filtering)
`get.categories.map(scad.data.frame)`	Get a list of user-defined category mappings (used for merging/redefining factor levels)
`validate.scad.data.frame(scad.data.frame)`	Auxiliary function (input validation)
`save.scad.data.frame(scad.data.frame)`	Auxiliary function (saving data frames)

4 How did these characteristics of the social conflict process change after the beginning of the AS?

4.1.1 *A Note on Geographic Information, Accuracy of the Information, and Exceptional Events*

In the SCAD database each record (row) corresponds to an event in a particular location. Events that took place in more than one location are listed separately. If the user is not interested in detailed geographic information about the events (as is the case in the present work), the database must be filtered to avoid over-counting the same event, as described in [63]. The R function `get.events.in.database()` implements this procedure. Events resulting from escalation are not coded in separate records, and detailed information about escalated events (e.g., the estimated number of participants) is not be available in SCAD. In the analysis below, only primary events will be taken into account.

As stated in the SCAD codebook, it is often difficult to obtain information about conflict events from published news accounts [63]. For instance, the estimated number of participants is given in terms of its order of magnitude and coded as a categorical variable (`npart`). For a substantial number of events, even the order of magnitude is unknown. This poses significant limitations for comparing the size of conflict activity peaks in ABM simulations with the corresponding values in real events. If the variable of interest is the proportion of a country's population involved in conflict events instead of the estimated number of participants, the analysis is further complicated by the differences of population size and demographic growth among the countries analyzed. For events with very large number of participants or any other exceptional characteristic (e.g., very long duration) it is also important to analyze the contents of `issuenote` and other fields (e.g., `etype`, `duration`, `escalation`, `actor1`, `location`) in order to interpret their significance and relationship with other events.

In some cases, the start and end dates could not be determined precisely [63]. For some events, the coded value of `duration` is several hundred days. Analysis of the summary description in the `issuenote` field shows that these event records usually correspond to an aggregate of many shorter but related events. Thus, in the statistical analyses of duration and recurrence time, it is necessary to inspect these exceptional events more carefully.

4.1.2 *Question 1: How Important Were Demonstrations and Riots, in Terms of Number of Events and Estimated Number of Participants?*

The analysis of this question started by selecting all events that were targeted against central governments for the countries of interest. The date December 15th, 2010—mid-December, two days before Bouazizi's self-immolation and three days before the first subsequent protests—was adopted in this work as conventional date from splitting the periods before and after the AS.

Table 4.3 shows the number and percentage of conflict events by type and Fig. 4.1 the number of conflict events of each type by estimated number of participants, before and after the beginning of the AS. The frequency increased notably after the start of the AS, from an average of 0.41 to 7.21 events per week (more than one event per day, considering all event types and sizes in the eight countries taken together). Demonstrations were clearly the most common form of conflict event and also the most important in terms of estimated number of participants. It is interesting

Table 4.3 Number and % of conflict events by type in Algeria, Egypt, Libya, Mauritania, Morocco, Sudan, and Tunisia, before and after 2010-12-15 (Source: author, based on [71])

	Before		After	
Demonstration	360	(60.7%)	601	(52.5%)
Riot	99	(16.7%)	111	(9.7%)
Strike	72	(12.1%)	71	(6.2%)
Anti-governmental violence	46	(7.8%)	342	(29.9%)
other	16	(2.7%)	19	(1.7%)

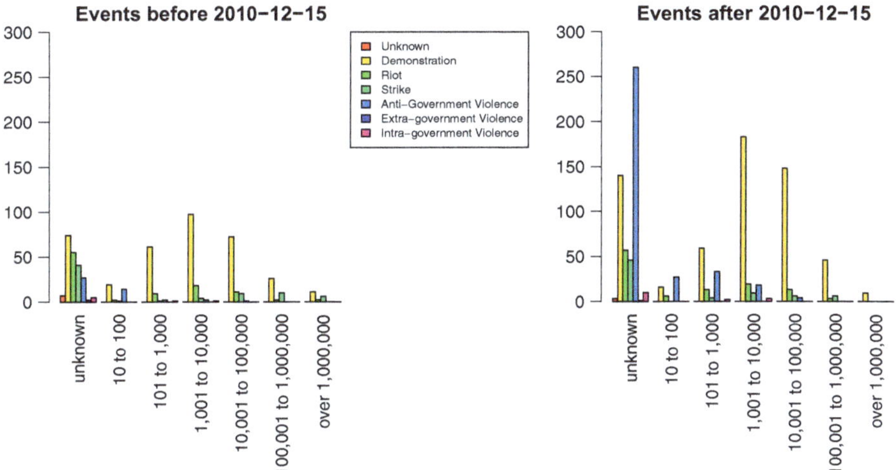

Fig. 4.1 Type of social conflict events against central governments in Algeria, Egypt, Libya, Mauritania, Morocco, Sudan, and Tunisia, before and after 2010-12-15, by estimated number of participants (Source: author, based on [71])

to note that very large demonstrations and strikes (with more than 100,000 and more that 1,000,000 estimated participants) occurred before December 15th, 2010 (Fig. 4.1), showing that strong signs of social conflict were already manifest in the countries analyzed, years before the beginning of the AS. Riots were of secondary importance when compared with demonstrations. Peaceful conflict manifestations, either organized (demonstrations and strikes) or spontaneous (demonstrations), were much more frequent and mobilizing than events involving violence, which is in line with Gene Sharp's theory of non-violent action.

Figure 4.1 also shows that the order of magnitude of the estimated number of participants is not known for a substantial proportion of events.[2] However, it is reasonable to assume that events with unknown number of participants were

[2]More specifically, this is the case for 728 events, representing 42% of all events in the database from May 17th, 1983 to December 31st, 2013.

not salient, otherwise estimates would be given. Therefore, events with unknown estimated number of participants were excluded from all the analyses below.

4.1.3 *Question 2: Which Were the Issues, Organization, and Escalation in Large Demonstrations and Riots?*

Large events involving a significant proportion of a country's population and the issues that triggered such events are of primary interest for the present work. Before proceeding, it is necessary to define the lower limit of estimated number of participants for a "large" event. It is plausible to consider that limit to be 1% of a country's population. Taking 10,000,000 as the order of magnitude of the latter, the events of interest for this question are those with 100,001 to 1,000,000 and with more than 1,000,000 estimated participants, which from here on will be conventionally called "very large" and "huge" events, respectively. For the time period of recorded events (before and after the AS) and the eight countries analyzed, there were 92 demonstrations and only seven riots that were very large or huge.

Figure 4.2 shows the number of very large and huge demonstrations for each issue category. The main issue was `human rights, democracy`, particularly after the beginning of the AS. The main issues causing very large and huge riots

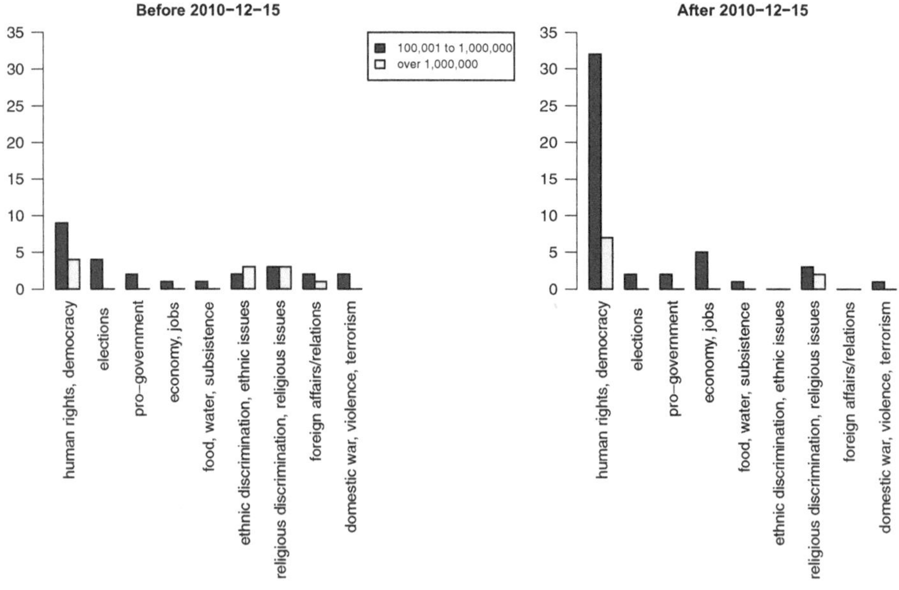

Fig. 4.2 Issues in demonstrations with more than 100,000 estimated participants in Algeria, Egypt, Libya, Mauritania, Morocco, Sudan, and Tunisia, before and after 2010-12-15 (Source: author, based on [71])

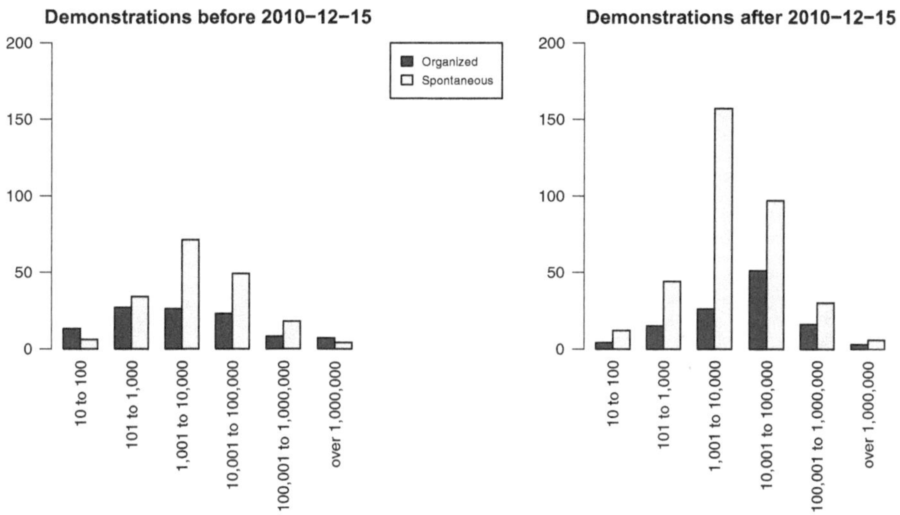

Fig. 4.3 Organized vs spontaneous demonstrations with known estimate of the number participants in Algeria, Egypt, Libya, Mauritania, Morocco, Sudan, and Tunisia, before and after 2010-12-15 (Source: author, based on [71])

(not shown) were more heterogeneous [45]. Clearly, the main drive of the AS was the strive of the populations for more democracy and human rights (individual liberties and political participation), and large peaceful demonstrations were the tool for pressing central governments. Figure 4.3 shows the number of organized and spontaneous demonstrations for which the estimated number of participants is known, including very large and huge demonstrations, before and after the beginning of the AS. The observed proportion of spontaneous demonstrations for all size categories, even if convoked or publicized by activists, is a strong argument supporting the study of large-scale conflict from the viewpoint of complexity. The proportion of spontaneous demonstrations increased after the beginning of the AS. The distribution of the number of spontaneous riots (not shown) is nearly symmetric like the one shown in Fig. 4.3, but the proportion of spontaneous riots was higher than for demonstrations [45] .

It is also important to consider escalation, to analyze the transitions from peaceful to violent forms of collective behavior.[3] Figure 4.4 shows the distribution of escalation in very large and huge demonstrations before and after the AS. Escalation to spontaneous riots was very significant, in particular for huge demonstrations. After the beginning of the AS, the number of huge demonstrations that escalated to spontaneous violent riots was larger than the number of huge demonstrations

[3]In SCAD, "escalation" has a different meaning than is usual. It refers to events of one type that degenerated into another type, regardless of violence intensity, organization, etc. It does not mean that the degenerated event was more violent than the original one.

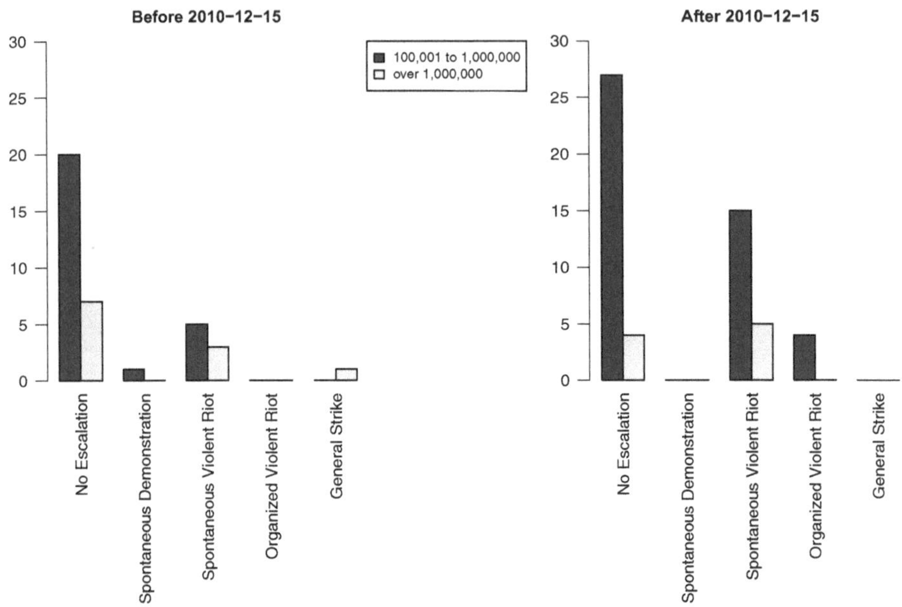

Fig. 4.4 Escalation in very large and huge demonstrations in African "Arab Spring" countries, before and after the beginning of the AS (Source: author, based on [71])

that remained peaceful. In general, the proportion of participants that rioted in each case is not indicated, but some `issuenote` descriptions mention how many people were injured or killed.[4]

[4]Analyzing the `issuenote` field for all events in the eight countries would be extremely time consuming, but is feasible for very large and huge events and sometimes yields useful details. For instance, the `issuenote` description for the very large demonstration in Algiers, Algeria, on January 22nd, 2011, whose main issue was `food, water, subsistence` reads: "Arab Spring. Rock and chair throwing protesters defy a ban on public gatherings and clash with police, they also demand radical change to the regime. 20,000 police stop the protests, leaving 5 dead and 800 hurt." This shows that although the coded `issue1` was related to economic RD, the event was in fact associated with the AS movement and its main issues (struggle for individual liberties and regime change), and was violently repressed by a massive police force. The description of the huge demonstration in several cities in Egypt on January 25th, 2013, whose main issue was `human rights, democracy` reads: "An estimated 500,000 people marched across Egypt against the Morsi government and against the Muslim Brotherhood and the protests turned violent, with 11 people killed and several hundred wounded." This also illustrates the uncertainty of the estimates of participation (the `npart` field is coded "over 1,000,000," which is not consistent with the estimate in the `issuenote`). For the very large, nationwide demonstration in Morocco on February 2nd, 2011, whose main issue was `economy, jobs`, the description reads: "Arab Spring. Protests erupt in favor of constitutional reform, social justice, and economic reform. 37,000 people take to the streets. Marches escalate to violence in Hoceima, where several people die after setting a bank on fire." Once again, the number of participants mentioned in the `issuenote` does not match the `npart` value.

4.1.4 *Question 3: What Were the Patterns of Recurrence, Duration and Size of Demonstrations and Riots?*

In the analysis of this question, the purpose was to examine the distributions of event size, duration and recurrence (time between successive events) for each country separately. Before proceeding, it is necessary to define the lowest size of demonstrations and riots considered relevant for the analysis of this question. This is determined by three factors: the order of magnitude of the population size for the countries considered, the distributions of estimated number of participants (`npart`), and the manageable number of agents that can be included in ABM.

Analysis of the SCAD database shows that the most frequent value of the `npart` categorical variable is `1,001 to 10,000` for five countries before the beginning of the AS and for six thereafter [45], which corresponds to 0.01% of the order of magnitude of the population size, or to one "active" citizen in a simulation with 10,000 agents. Thus, it is reasonable to take 1,000 estimated participants as the lowest size of demonstrations relevant for the analysis of this question. This estimate is also plausible for the case of riots [45].

* * *

Figure 4.5 shows box-and-whisker diagrams for the duration of demonstrations and riots with more than 1,000 estimated participants for the African AS countries. Before the beginning of the AS the median of duration is two days for Egypt and Tunisia and one day for the other countries. The third quartile of duration is 7.75 for Egypt, 10 for Mali and 13.5 for Morocco, and less than or equal to two for the other

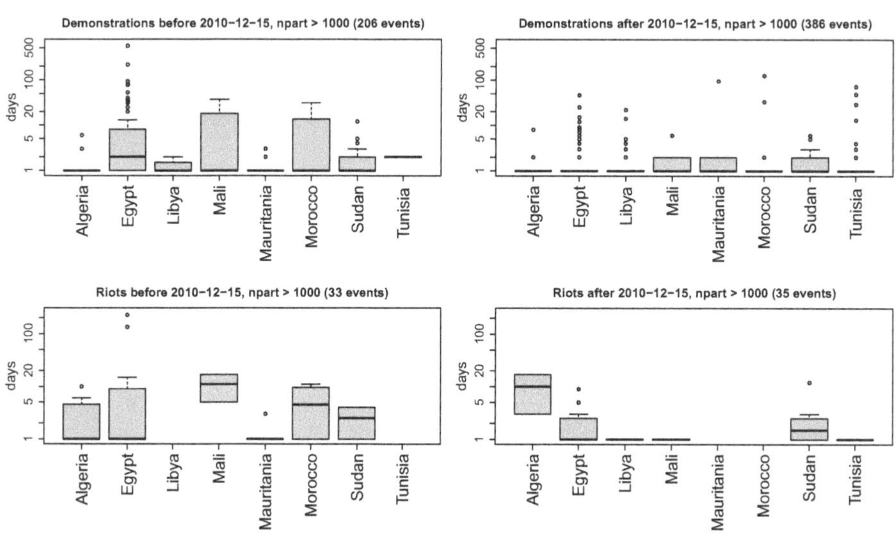

Fig. 4.5 Box-and-whisker diagrams for the distribution of duration of demonstrations and riots with more than 1,000 estimated participants in Algeria, Egypt, Libya, Mauritania, Morocco, Sudan, and Tunisia (Source: author, based on [71])

countries. Thus, except for Egypt, Mali and Morocco, about 75% of the demonstrations with more than 1,000 estimated participants were short events. Egypt is notable for the exceptionally long duration of some demonstrations. Analysis of the `issuenote` field shows that these were in fact long successions of shorter events related to particular conflict situations, coded as a single record for convenience or lack of more precise information [45]. After the beginning of the AS the median of duration is one for all countries; the third quartile is one for Algeria, Egypt, Libya, Morocco, and Tunisia, two for Sudan and 1.75 for Mali and Mauritania.

Although demonstrations tended to be shorter after the beginning of the AS, Fig. 4.5 also shows that more countries had demonstrations with exceptionally long duration than before the AS.[5]

Riots with more than 1,000 participants were in much smaller number than demonstrations and the estimates for the median of duration are more heterogeneous than for demonstrations. Riots tended to be shorter after the beginning of the AS for Egypt, Mali, and Sudan, but the reverse is true for Algeria.

* * *

Figure 4.6 shows box-and-whisker diagrams for the time interval between successive demonstrations and riots with more than 1,000 estimated participants for the African AS countries.

Before the beginning of the AS demonstrations were already frequent in Algeria, Egypt, and Mauritania (the median of the time interval is 18, 29, and 32 days respectively, with 75% of time intervals smaller than half a year for Algeria and Egypt). After the beginning of the AS demonstrations became more frequent than in the previous period for all countries except Mauritania, which is consistent with the notable increase of the number of conflict events. This is particularly striking for Egypt, Libya, Mali, Morocco, Sudan and Tunisia.[6]

Riots were rarer than demonstrations and consequently the time intervals between successive events was larger than for demonstrations, and their distributions are also heterogeneous among the countries analyzed.

---

[5]The longest demonstration after the beginning of the AS started in Morocco on January 1st, 2012, with `duration` = 126 days, `npart` = "10,001 to 100,000" and its `issuenote` is as follows: "Thousand of Berbers participated in protests throughout various regions of Morocco to demand the government for more democratic reform and more government services for Berbers." In Mauritania, the demonstration that started on May 2nd, 2012, with `duration` = 95 days, `npart` = "10,001 to 100,000" has the `issuenote`: "The Coordination of a Democratic Opposition, a coalition of opposition parties, held a lengthy series of weekly protests in Nouakchott demanding the resignation of President Mohamed Ould Abdel Aziz. Police arrested, beat, and tear gassed protesters on several occasions. Some youth demonstrators resorted to throwing rocks and burning tires. Attendance ranged between hundreds and thousands." In Tunisia, the demonstration that started on September 17th, 2012, with `duration` = 71 days, `npart` = "1,001 to 10,000" has the `issuenote`: "Prisoners protested against the Justice Ministry." In Egypt, the longest demonstration after the beginning of the AS had `duration` = 46 days.

[6]The median of the interval between successive events changed from 18 to 12.5 days for Algeria; [29 to three days for Egypt]; 1,801.5 to 18.5 days for Libya; 266 to 36 days for Mali; 346.5 to 15 days for Morocco; 226 to 21 for Sudan; and in Tunisia the median of the time between successive demonstrations was 12 days after the beginning of the AS.

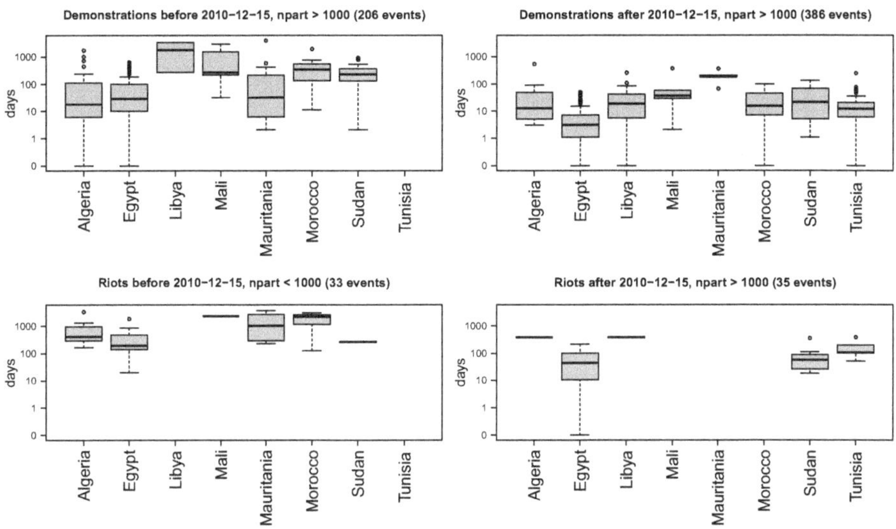

Fig. 4.6 Box-and-whisker diagrams for the distribution of the time interval between of demonstrations and riots with more than 1,000 estimated participants in Algeria, Egypt, Libya, Mauritania, Morocco, Sudan, and Tunisia (Source: author, based on [71])

Another variable of interest for comparing records of conflict event with ABM simulations is the proportion (or %) of the time in which a country or set of countries had calm (no conflict events) or activity (at least one conflict event taking place).[7] Figure 4.7 shows the percentages of the time with calm and active events for demonstrations and riots with more than 1,000 estimated participants, considering each country individually, and all eight countries taken together.

Before the beginning of the AS, Egypt was the only country with a significant percentage of the time (13.79%) with demonstrations occurring. In the other countries, the percentage of the time with demonstrations occurring was very small, so that the corresponding percentage for all countries together (16.27%) is hardly distinguishable from that for Egypt in Fig. 4.7.[8] After the beginning of the AS, the situation changed dramatically. The percentage of time with demonstrations for all countries jumped to 56.6%.[9] The proportion for Egypt was 26.33%, and countries

[7]This proportion or % is independent of the time unit and thus of the time step in an ABM, which for models of "abstract" type like Epstein's is usually indefinite.

[8]The second and third countries were Morocco with 1.09% and Mali with 0.72% of the time with active demonstrations. The difference between the sum of the countries' individual % of time with active demonstrations (16.96%) and aggregated value (16.27%), is a measure of the percentage of the total time with demonstrations occurring simultaneously in more than one country.

[9]This is a surprising result, especially because only major demonstrations against the central governments were considered, which in itself gives an objective measure of the intensity of social conflict manifest in the AS.

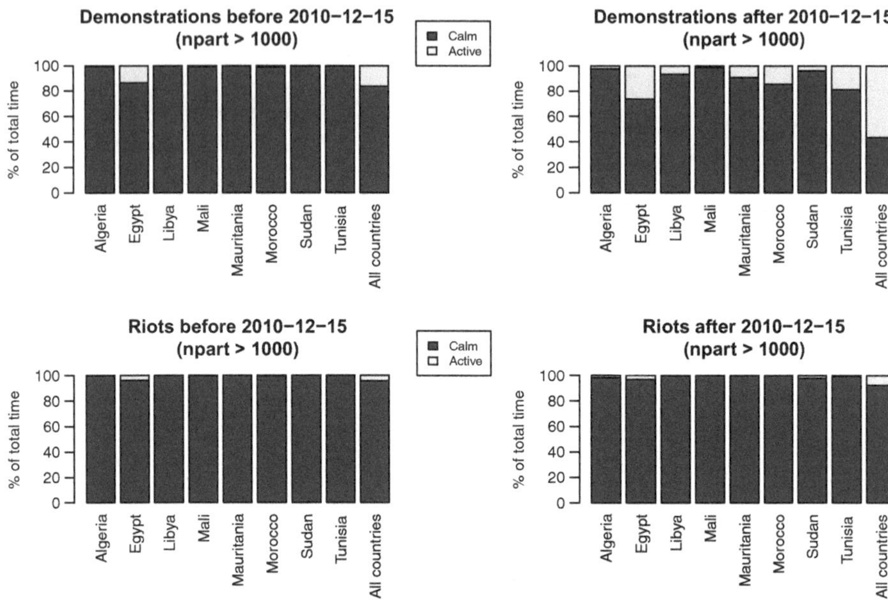

Fig. 4.7 Proportions of time with calm and activity in demonstrations and riots with more than 1,000 estimated participants for African AS countries, before and after the beginning of the AS (Source: author, based on [71])

which previously had only residual % of total time with active demonstrations had much higher %'s after the beginning of the AS: Tunisia 18.69% (previously 0.02%), Morocco 14.47%, Mauritania 9.07%, Libya 6.65%, and values below 5% for Algeria, Mali, and Sudan.

Riots were rarer events than demonstrations, and consequently the % of time with active riots was smaller than for demonstrations. The value for all countries taken together was 4.39% before and 7.46% after the beginning of the AS. Egypt was the only country with significant % of time with active riots (3.87% before and 3.14% after the beginning of the AS).

* * *

To conclude the exploration of SCAD with respect to question 3 above, it remains to analyze the time history of event size for demonstrations and riots with more than 1,000 estimated participants. The two variables of interest are the event size (estimated number of participants), and the proportion of the population involved in conflict events. The former is of primary importance in terms of news impact and international visibility. The latter is more relevant as a measure of a country's mobilization and also more useful for comparing simulation results, for it is independent of the population size or the number of agents used in an ABM.

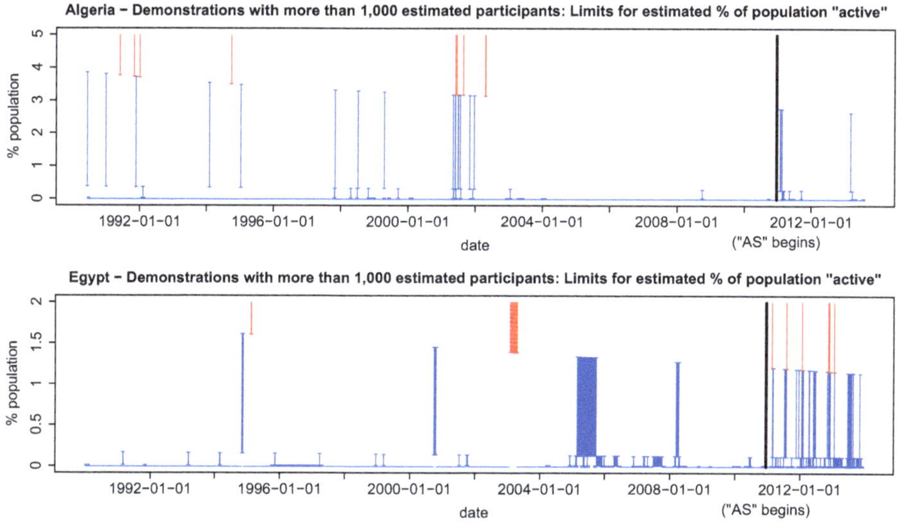

Fig. 4.8 Time history of the range of estimated % of the population participating in demonstrations with more than 1,000 estimated participants, for Algeria and Egypt. Events with more than 1,000,000 estimated participants are represented by *dashed red lines* with indefinite upper limit. All other events are represented by *blue error bars* (Source: author, based on [71] and [75])

As mentioned before, only the order of magnitude of the estimated number of participants is provided as a categorical variable in the SCAD database. There is no general theoretical or empirical procedure for obtaining an expected value for the number of participants within each category. Therefore, it is only possible to construct time histories for the estimated range of event size. To obtain the ranges for the proportion of the population involved in conflict events, it is necessary to divide the size limits for each category by the country's population. In the analysis below, this was done by dividing the size limits by the value of the country's population at the start date of each event, computed via a spline interpolation of the UN World Population Data [75]. For huge events with `npart = "more than 1,000,000"`, the lower limit (1,000,000) was divided by the interpolated estimate of the country's population, and the upper limit was left undetermined.

Using this approach, graphical representations of the time history of the estimated range of % of population participating in conflict events were obtained for demonstrations and riots with more than 1,000 estimated participants, for the eight countries. Only the cases of demonstrations in Algeria, Egypt, Morocco and Tunisia will be presented, since they are the most representative and interesting [45].

Figures 4.8 and 4.9 show the time history of the estimated ranges of the % of population involved in demonstrations for Algeria and Egypt, and for Morocco and Tunisia, respectively. In these figures, the range of estimated % of the population is represented by blue error bars drawn in solid line, except for demonstrations with more the 1,000,000 estimated participants, which are represented by dashed red

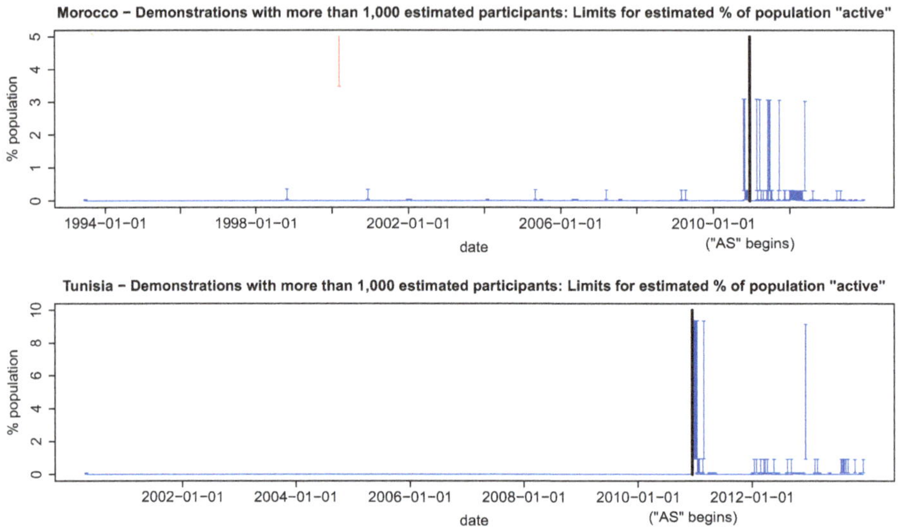

Fig. 4.9 Time history of the range of estimated % of the population participating in demonstrations with more than 1,000 estimated participants, for Morocco and Tunisia. Events with more than 1,000,000 estimated participants are represented by *dashed red lines* with indefinite upper limit. All other events are represented by *blue error bars* (Source: author, based on [71] and [75])

lines with indefinite upper limit. The conventional date December 15th, 2010 is marked by a thick black line in each graph.

Figure 4.8 shows that in Algeria and Egypt, which are very populous countries, the estimated % of the population involved in demonstrations was below 5%.[10] Plausible ranges of the % of the population for very large demonstrations are [0.5,4.0] for Algeria and [0.2,1.6] for Egypt. These proportions were possibly exceeded in the few huge demonstrations shown in this figure. The effect of the AS on the time history of demonstrations was different in these two countries: in Algeria, the frequency of conflict events was not significantly altered by the AS, whereas in Egypt demonstrations (both very large and huge) became more frequent after the beginning of the AS (confirming the results in Figs. 4.5, 4.6 and 4.7). The thick blue bars in the plot for Egypt correspond to demonstrations with exceptional duration (also shown as outliers in Fig. 4.5).

The situation was somewhat different in the cases of Morocco and Tunisia (Fig. 4.9). Except for one isolated huge demonstration, there were no massive demonstrations in Morocco before the beginning of the AS. In Tunisia, there were no major demonstrations before the beginning of the AS. Just before December

[10]The monotonic decrease of the width of the error bars, particularly for very large demonstrations (`npart = "100,000 to 1,000,000"`) is due to the effect of population growth [45], and consequently is more pronounced for Egypt.

15th, 2010, there were two demonstrations (one of them very large) in Morocco.[11] In Tunisia, there were very large demonstrations after that date, but they were less frequent than in Egypt for the same period. Plausible ranges of the % of the population for very large demonstrations are [0.4,3.0] for Morocco and [1.1,8.5] for Tunisia.

4.2 Analysis of Fragile States Index Indicators for the African "Arab Spring" Countries

The Fragile States Index (FSI), issued by The Fund for Peace (FFP) since 2005, is an annual index for measuring vulnerability to conflict that includes 178 countries. The FSI is computed from a large number (about 11.000) of public sources using the Conflict Assessment System Tool (CAST) software [5], based on twelve political, social and economic indicators. Some of these indicators, particularly "Legitimacy of the State" and "Human Rights and Rule of Law," are important in the context of the present work.

Before analyzing any scores for the countries of interest, an exploratory factor analysis (EFA) was done to determine whether or not the FSI indicators could eventually be reduced to a smaller number of meaningful constructs (factors). Figure 4.10 shows the correlation matrix of the FSI scores for the twelve FSI indicators, considering the 178 countries and years 2006 to 2015 (1753 records). All scores are positively correlated, with all Pearson correlation coefficients larger than 0.5.

After visual inspection of the correlation structure of the FSI scores, a factor analysis was done in the following steps: (*i*) estimation of the optimal number of factors to extract; (*ii*) computation of factor solutions for the number of factors obtained in the previous step as well as for one more and one less factor ("bracketing"); (*iii*) inspection of the factor solutions for fit indices, communality and loadings of the variables; and (*iv*) selection of the solution with best fit indices

[11]The time line of demonstrations for Morocco in Fig. 4.9 induces the impression that the AS started in Morocco, which was not the case. The first of these two events was a spontaneous demonstration by Saharawis, with issue `"economy, jobs"` and an estimated number of participants between 100,001 and 1,000,000, which started on October 18th, 2010 and ended on November 10th, 2010. The `issuenote` for this demonstration reads: "Western Saharans set up a camp to protest, demanding a better standard of living and more jobs." The second event was a spontaneous demonstration by Textile Workers, with issue `"economy, jobs"`, with an estimated number of participants between 10,001 and 100,001, which started on October 29th, 2010 and ended on November 27th, 2011. The `issuenote` is: "Textile workers stage a sit-in, protesting the decision to close four textile plants." The two following demonstrations in Morocco started on February 10th and 20th, 2011, and were related to the AS. The `issuenote` field for these events reads: "Arab Spring. A Moroccan dies after self-immolating to protest his situation after being dismissed from the army" and "Arab Spring. Protests erupt in favor of constitutional reform, social justice, and economic reform. 37,000 people take to the streets. Marches escalate to violence in Hoceima, where several people die after setting a bank on fire." The `issue1` field for these two events is coded as `economy, jobs`. These remarks show the importance of analyzing the details of each situation when exploring the SCAD database.

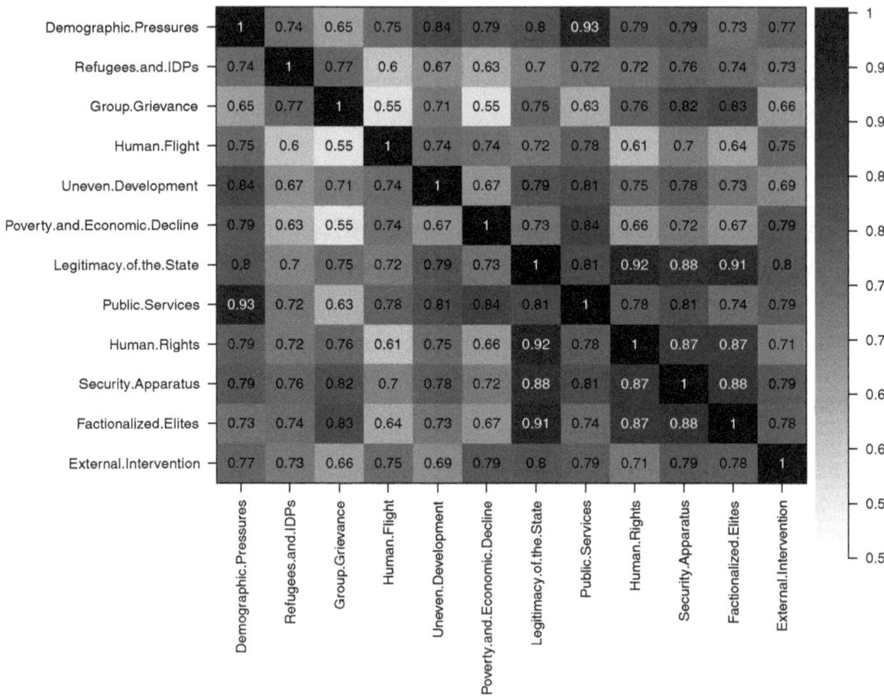

Fig. 4.10 Matrix of the Pearson correlation coefficients for the twelve FSI indicators. Fit indices: `RMSEA` = 0.126, `TLI` = 0.938, and `BIC` = 506.1 (Source: author, based on [70])

that is theoretically meaningful. The factor analysis was done in R using the `psych` library [59]. A general description of the method for doing EFA can be found in [39].

The number of factors to extract was estimated by means of parallel analysis performed using the `psych::fa.parallel`. The factor analyses were ran using the `psych::fa` function, with factoring method "generalized least squares," rotation method was "oblimin" and 2000 bootstrap iterations. The estimation of the factors' reliability was done using the `psych::alpha` function. Figure 4.11 shows the factor diagram for the solution with best fit indices. The `RMSEA` and `TLI` fit indices for this solution are 0.126 and 0.938, respectively.[12] These values indicate poor fit. Also, the number of variables loading on factors `GLS2` and `GLS4` is insufficient [39]. Nevertheless, this solution suggests that vulnerability to conflict, as described by the FSI indicators, can be associated with degrading economy and demographic pressures, the political system, internal group grievances and refugees, and external interventions.

[12] `RMSEA`—Root Mean Square of the Approximation; `TLI`—Tucker-Lewis Index of factoring reliability. `RMSEA` is an absolute fit index, whereas `TLI` is a relative fit index. A description of these (and other) fit indices and their boundaries for good, acceptable and poor fits can be found in [39] and [59].

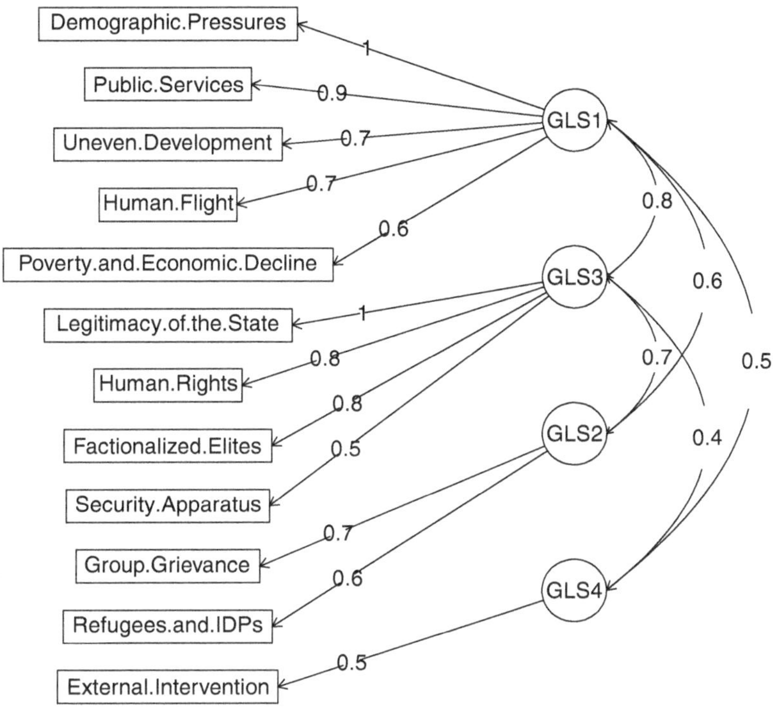

Fig. 4.11 Factor diagram for the four-factor solution of the scores of the twelve FSI indicators (Source: author, based on [70])

Figure 4.12 shows the FSI scores of the "Legitimacy of the State" for the countries of interest. The values of the FSI indicators and the final scores are evaluated in a 0–10 scale from "best" to "worst," but in view of setting plausible values for global variables in ABM it is convenient to express them in a 0–1 scale from "worst" to "best," via a simple linear mapping.

It is not the goal of the present work to discuss whether or not the FSI is a good proxy for evaluating the legitimacy of political systems, or to draw considerations about the legitimacy of the governments of the selected countries before or after the AS. For the purpose of model exploration, the important observation is that in a 0–1 scale the values of the "Legitimacy of the State" FSI indicator shown in Fig. 4.12 are significantly lower than those considered in previous studies of social conflict using ABM. Therefore, the relevant issue is to determine whether or not the proposed model can produce solutions with different regimes (calm, intermittent rebellion or permanent turmoil) for low values of legitimacy.

Figure 4.12 also shows that the FSI "Legitimacy of the State" indicator has no significant predictive value for anticipating phenomena like the AS. This can be explained by several factors. First, the scores are computed based on time-delayed data. The FSI legitimacy score does not take into account important aspects like

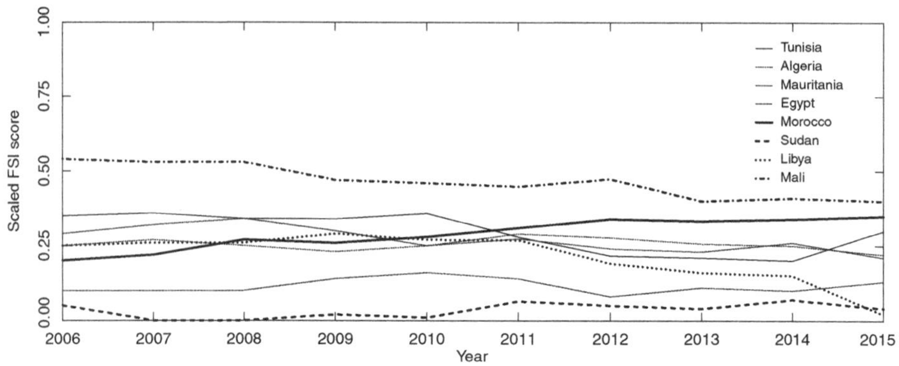

Fig. 4.12 Scaled FSI "Legitimacy of the State" scores for Algeria, Egypt, Libya, Mauritania, Morocco, Sudan, and Tunisia (Source: author, based on [70])

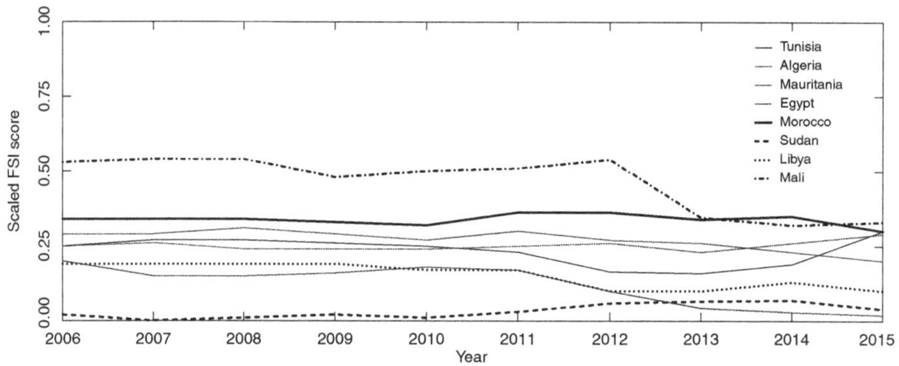

Fig. 4.13 Scaled FSI "Human Rights and Rule of Law" scores for Algeria, Egypt, Libya, Mauritania, Morocco, Sudan, and Tunisia (Source: author, based on [70])

the components of legitimacy (as in Gilley [33, 34]), or the difference between the legitimacy of the government and the legitimacy of the regime.[13] Also, the outburst of large-scale conflict processes depends on many factors not accounted by a single indicator, such as triggering events. It is also interesting to question whether or not legitimacy feedback effects were significant. The information in Fig. 4.12 does not lead to conclusive answers, although in the cases of Tunisia, Egypt, and Libya the legitimacy scores dropped after the beginning of the AS.

Figure 4.13 shows the scaled FSI "Human Rights and Rule of Law" scores for the same sets of countries of Fig. 4.12. The conclusions are same as for the legitimacy (Fig. 4.11 shows that they load on the same factor), i.e. the indicators show no significant variation before the uprisings and the values are inconsistent

[13]Bischof [9] argues that monarchies are more stable than republics in authoritarian regimes.

with the sequence and the intensity of conflict events in each country. With the notable exception of Tunisia, the scores show a generalized negative trend after the AS, indicating a (paradoxical) potential increase of political deprivation. Thus, the "Legitimacy of the State" and "Human Rights and Rule of Law" indicators are essentially equivalent in terms of the information they convey about the relationship between different countries and the time variation of the respective scores.

4.3 Analysis of the Freedom in the World Indicator for the African "Arab Spring" Countries

Since the main issues in the AS were political (demand for democracy and human rights), it is interesting to compare the FSI "Human Rights and Rule of Law" score with the "Freedom in the World" score, available from the Freedom House datasets.

Figure 4.14 shows the scaled "Freedom Rating" scores for the same countries considered before. Before the beginning of the AS the scores are relatively stable and the ordering of the countries is not very different from that of the FSI "Human Rights and Rule of Law" indicator. After the AS, the results of the two indicators are somewhat different: (*i*) Tunisia occupies the first place and is considered "Free;" (*ii*) Mali and Morocco come second and third and are "Partly Free" (with Mali worsening after 2012 from "Free" to "Partly Free"); (*iii*) Libya and Egypt improved from 2011 to 2013, but worsened again since 2014. In both indicators, Sudan's score is the worst.

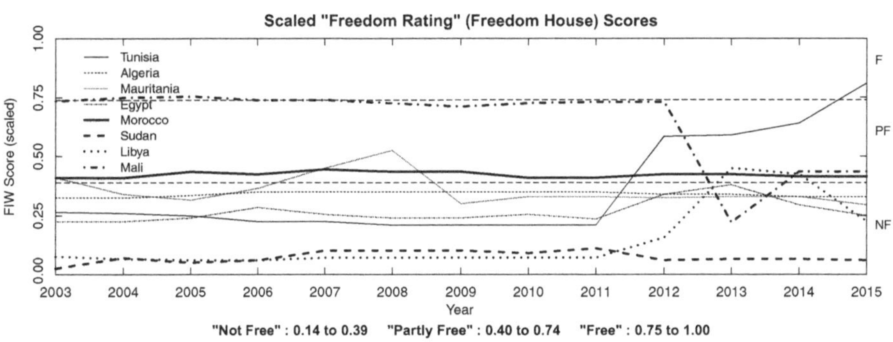

Fig. 4.14 Scaled "Freedom in the World" scores for Algeria, Egypt, Libya, Mauritania, Morocco, Sudan, and Tunisia (Source: author, based on [28])

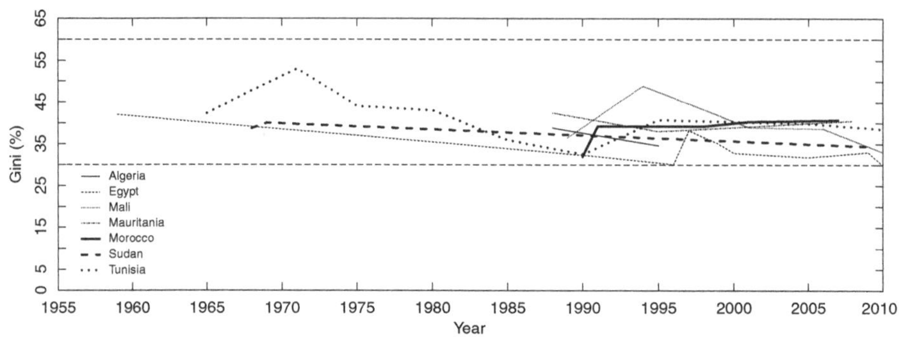

Fig. 4.15 "All the Ginis" scores of the Gini index for "Arab Spring" countries in Africa. The generally assumed bounds for "small" and "large" inequality (30% and 60%, respectively) are represented by horizontal dashed lines (Source: author, based on [49])

4.4 Analysis of the "All the Ginis" Dataset for the African "Arab Spring" Countries

Although the main issues in the AS were not due to economic RD, it is useful to consider indicators of welfare inequality to characterize the potential relevance of economic RD in the African AS countries. For this purpose, it is worthwhile analyzing the values of this index available in the "All the Ginis" dataset [49].

Figure 4.15 shows the aggregated Gini index for the countries of interest. The time coverage is very uneven in different countries and ends before the beginning of the AS for all countries. Thus, trends in inequality and their potential effect of the AS cannot be inferred from the "All the Ginis" dataset. Nevertheless, it is plausible to assume that welfare inequality was moderate (closer to 30% than to 60%) and relatively uniform among the countries considered. This suggests that 0.4 (or 40%) can be adopted as a representative value of the Gini index for the African AS countries.

4.5 Concluding Remarks

The analysis of the SCAD database led to the following answers to the questions stated above (page 30):

- Demonstrations were the most important form of conflict event, both in number of events and mobilization (estimated number of participants). Riots were rarer and less significant. Egypt was the country with greatest number of conflict events, before and after the AS;
- The main issue in massive demonstrations (more than 100,000 estimated participants) was `human rights, democracy`. Massive riots had more het-

erogeneous causes. Spontaneous demonstrations and riots were more numerous than organized events, which suggests that the methods of complex systems studies are useful for studying large-scale conflict processes. Escalation of large demonstrations to riots was significant, but riots did not escalate to more intense forms of violence;
- Demonstrations and riots occurred intermittently. In general, demonstrations were short events with median 1 day and 75% of the cases with duration less than ten days. The interval between successive events was heterogeneous among the eight countries, and was much shorter after the beginning of the AS (characteristic value of 15–20 days). The % of the population participating in demonstrations was smaller for more populous countries (e.g., Egypt and Algeria) than for less populous countries. Typical ranges of these % are [0.5,4.0] and [1,10], respectively, meaning that activity peaks involving 10% or more of the population are very unlikely;
- There were very large demonstrations and riots in Algeria, Morocco, and Egypt several years before December 2010, but after the beginning of the AS the frequency of events increased notably. Taking all eight countries together, the % of total time with demonstrations increased from 16% to 57%. For riots, the corresponding figures are 4% and 7%, showing that demonstrations were indeed the main form of protest after the beginning of the AS. The most common issue in massive demonstrations was `human rights, democracy` before and after the beginning of the AS. Demonstrations tended to have shorter duration (median 1 day) and the interval between successive events (15–20 days) after the beginning of the AS.

The analysis of the FSI, FWI and "All the Ginis" lead to the following general conclusions. First, none of these indicators had "prognostic" value for anticipating either the onset of the AS or the order by which the movement propagated. Also, the scaled FSI scores for "Legitimacy of the State" were below 0.4 (in a scale 0-1) for all countries analyzed except Mali, with very low values for Sudan and Egypt. In some countries, the scaled "Legitimacy of the State" and "Human Rights and Rule of Law" scores deteriorated after the beginning of the AS. This supports the conjecture that legitimacy feedback is significant, with characteristic variations in the range [20%, 40%]. Finally, a plausible value of the Gini index for the countries analyzed is 0.4 (or 40%).

Chapter 5
ABM of Civil Violence: ODD Description

This chapter contains a description of the proposed ABM, using the well-known ODD protocol [36]. The model was implemented in NetLogo [77] and is based on the same two types of agents—citizens and law-enforcement—as in Epstein's model, but "citizen" agents have one subtype ("activist") used for implementing network influence effects. The salient aspects of the model description are centered on the submodels for the risk perception, the expression of hardship in terms of value-sensitive RD and endogenous legitimacy feedback, which provide extensions to Epstein's model. The model includes two networks called `group` and `infl` (for "influentials"), to represent two different types of social influence (due to small and highly cohesive groups or "activists," respectively). Network influences were implemented in the model using a form of dispositional contagion [23]. The chapter also includes an outline of the R scripts for pre- and post-processing the model results, using the RNetLogo package [72, 73].

5.1 Purpose

The purpose of the model is to simulate large-scale conflict processes against a central authority, including relative deprivation as a factor of political grievance and the mechanisms of "massive fear loss" (generalized vanishing of the risk perception), endogenous legitimacy feedback, and network influence effects.

C.M. Lemos, *Agent-Based Modeling of Social Conflict*, SpringerBriefs in Complexity, https://doi.org/10.1007/978-3-319-67050-8_5

5.2 Entities, State Variables, and Scales

The model was implemented in NetLogo [77], using the *Rebellion* model as a starting template [78], and has four entities: "observer," agents, networks, and environment (consisting of a grid of cells, or patches in NetLogo's terminology). The agents, networks, and model space will be described in this section. The "observer" is a special entity in the NetLogo system which represents the model user.

5.2.1 Agents

There are two types of agents, "citizens" and "cops." "Citizens" represent the population and may actively contest the central authority or not. "Cops" are the law enforcing officers and try to keep the order by arresting "active" citizens. Both "citizens" and "cops" have one move and one action rule.

Table 5.1 shows the attributes for "citizen" agents. There are two subtypes of citizens, "normal" and "activist," which are defined using the default `turtle` attribute `label`. "Citizen" agents can be in one of three states, "quiet," "active" (or "rebellious"), or "jailed." "Activist" citizens are the hubs of the `infl` network, as described below. They are defined only if the input variable `INFL?` is set to *true*. "Citizens" that are not "jailed" move and change state between "quiet" and "active" according to their move and action rules. "Activist" citizens (if present) differ from "normal" citizens only in their move rule.

The default move rule for "citizen" agents is the same as in Epstein's ABM:

Rule M$_1$: **if** $\texttt{jail}-\texttt{term} = 0 \wedge \mathscr{E}_v \neq \emptyset$ **then** move to a random cell $c_r \in \mathscr{E}_v$
else stand still

where $\mathscr{E}_v$ is the set of empty cells whose center lies within the agent's vision radius.[1] The move rule for "activist" citizens is:

Rule M$_2$: **if** $\texttt{jail}-\texttt{term} = 0 \wedge \mathscr{E}_v \neq \emptyset$ **then**
if $\mathscr{A}_v \neq \emptyset$ **then** move to one $c_r \in \mathscr{E}_v$ with $\texttt{min}(d(\hat{x}_g, \hat{y}_g))$
else move to a random cell $c_r \in \mathscr{E}_v$
else stand still

[1]A cell is considered empty if there is no agent inside it, or if it contains only jailed "citizen" agents. In the NetLogo implementation, "jailed" agents are hidden from view and do not interfere with the other agents' movement, but cannot be removed from the model space.

Table 5.1 Attributes of "citizen" agents (Source: author)

Variable name	Meaning	Description
`label`	Agent subtype	{ "normal","activist" }
`vision-a`	Vision radius v	"citizen" vision radius
`risk-aversion`	Risk-aversion R	$\sim \mathscr{U}(0,1)$
`perceived-hardship`	Hardship H	$=$ `RD` if `RD?` = *true*; $\sim \mathscr{U}(0,1)$ if `RD?` = *false*
`value`	Value capability	$\sim \mathscr{U}(0,1)$ if `RD-unif?` = *true*; Pareto-distributed if `RD-unif?` = *false*
`expectation`	Median of `value` of other visible "citizens"	Median of `value` of other visible "citizens"
`deprivation`	RD — a function of the difference between `expectation` and `value`	`RD` $=$ `max`$(0,$ `expectation` $-$ `value`$)^{\gamma}$ (see text)
`grievance`	Grievance G	$= RD \cdot (1 - L_p)$ if `RD?` = *true*; $= H \cdot (1 - L_p)$ if `RD?` = *false*
`social-influence`	Social influence	Weighted sum of dispositional contagion influences (see text)
`net-risk`	Net risk R	$\sim \mathscr{U}(0,1)$
`active?`	"citizen" state	*true* if "active"; *false* if "quiet" or "jailed"
`jail-term`	Jail term	$\sim \mathscr{U}(0, J_{\max})$
`visible-agents`	Agentset of other agents within vision radius	Used as a cache to speed up computations
`my-legitimacy`	List (vector) of perceived legitimacy, L_p	Perceived legitimacy in previous 1 to `L-memory` cycles

where $\mathscr{A}_v$ is the set of visible "citizens" that are "active" and $d(\hat{x}_g, \hat{y}_g)$ is the distance to the centroid $(\hat{x}_g, \hat{y}_g)$ [of] the set of patches occupied by "active" citizens. Thus, if "activists" see no "active" citizens, they follow the default move rule, otherwise they try to approach concentrations of rebellious citizens. This departure from random movement in "abstract" ABM is a simple representation of the "agenda setting bias" of "activists" and traditional media towards showing protests and violence.[2]

"Citizen" agents change state according to the threshold rule

Rule A: **if** $G - N + \mathscr{S} > T$ **then** be "active"
else be "quiet"

[2]The theoretical concepts on agenda setting can be found in [8].

where

$$G = H \cdot (1 - L_p) \text{ is the level of grievance;}$$
$$N = R \cdot P_a(\rho_v) \text{ is the net risk perception;}$$
$$\mathscr{S} = \texttt{w-group} \cdot \sum_{\mathscr{A}_k \in \texttt{AG}_i} (G_{\mathscr{A}_k} - N_{\mathscr{A}_k}) + \texttt{w-infl} \cdot \sum_{\mathscr{A}_l \in \texttt{AINFL}_i} (G_{\mathscr{A}_l} - N_{\mathscr{A}_l})$$
is the sum of network influences.

In these expressions, H is the hardship; L_p is the perceived legitimacy; R is the risk aversion; $P_a(\rho_v)$ is the estimated arrest probability; $\rho_v = (C_v/A_v)$, in which C_v is the number of visible "cops" and A_v is the number of "active" citizens visible to a generic "citizen;" $\mathscr{A}_i$; $\texttt{AG}_i$ and $\texttt{AINFL}_i$ are the sets of "active" citizens in the `group` network and "active-activist" citizens in the `infl` network, respectively, for citizen $\mathscr{A}_i$. The term $\mathscr{S}$ is zero if network influences are turned off.

"Cop" agents have one move rule and one action rule. The move rule for "cops" is:

Rule M_3 : **if** $\mathscr{E}_{v'} \neq \emptyset$ **then** move to a random cell $c_r \in \mathscr{E}_{v'}$
else stand still

where $\mathscr{E}_{v'}$ is the set of empty cells whose center lies within the cop's vision radius. The action rule for "cop" agents is:

Rule C : **if** $\mathscr{A}_{v'} \neq \emptyset$ **then**
select one random $\mathscr{A}_i \in \mathscr{A}_{v'}$
set "active?"$_{\mathscr{A}_i}$ = *false*
set jail-term$_{\mathscr{A}_i} = J \sim \mathscr{U}(0, J_{\max})$
move to $(x_{\mathscr{A}_i}, y_{\mathscr{A}_i})$
endif

where $\mathscr{A}_{v'}$ is the set of visible "active" citizens and $\mathscr{A}_i$ is a random "citizen" in this set.

5.2.2 Networks

The model includes two networks called `group` and `infl` (for "influentials"), which are intended to represent two different types of social influence. The former is related to strong influence in small and highly cohesive social groups (e.g., a family or clique of schoolmates), and the latter to influence of "activists" (which are influential agents) in a society. In the present version of the model the two networks do not change during the whole simulation.

The `group` network is set up by forming cliques of undirected links of type `group-member` between "citizens." The clique (group) size is defined via the `group-size` input parameter (Table 5.4). The `infl` network is set by connecting each "activist" (randomly chosen "citizen") to a % of the population defined by the `infl-size` input parameter, via directed links of `infl-follower` type. The `infl` network is a union of `num-infl` directed star networks, each with one "activist" as central hub. One "citizen" can be connected to more than one "activist" agent.

5.2.3 *Environment*

The scenario is a 2D homogeneous torus space, which combined with the random movement of ("normal") "citizen" and "cop" agents ensures that the probability of interaction between these two populations is independent of the position (i.e., there is no clustering or formation of "sanctuaries").

5.3 Process Overview and Scheduling

The model is implemented in two main procedures, `setup` and `go`, which initialize a new run and implement the main cycle, respectively. The `setup` procedure clears all variables from the previous run, resets the simulation clock (`ticks`), initializes the global variables, creates the agents and sets their attributes, builds the networks (`group` if `GROUP?` = *true* and `infl` if `INFL?` = *true*, respectively), and displays the simulation space. The `go` procedure consists of the following steps:

1. Activate all agents except "jailed" citizens by random order and execute their move and action rules;
2. For all "jailed" agents, decrement the `jail-term` variable by one. If $\texttt{jail} - \texttt{term} = 0$, "release" the "jailed" agent by moving it to an empty cell in the model space, setting its state as "quiet," and making it visible;
3. If `LF?` = *true* update the global legitimacy (see "Submodels" below for the details);
4. Advance the simulation clock;
5. Display the simulation space, and update the plots and monitors of the NetLogo interface.

For "cops," the first step is a straightforward application of Rules M_3 and C described above. For "citizens," the application of Rule A requires the following operations:

- Scan the environment and cache the visible "citizens"[3];
- If `LF?` = *true* update the perceived legitimacy L_p;
- If `RD?` = *true* set the hardship as the relative deprivation (RD);
- If `GROUP?` = *true* and/or `INFL?` = *true*, compute the social influence term $\mathscr{S}$;
- Compute G and N;
- Update the state according to the action Rule A.

5.4 Design Concepts

5.4.1 *Basic Principles*

The basic principles used in the development of the ABM are:

- Preserve the basic simplicity and "minimalist generative capacity" of Epstein's original ABM (i.e., keep the same types of agents and simple threshold action rule for "citizens");
- Model network influences via the mechanism of *dispositional contagion* instead of *behavioral imitation* (as in [23]), considering that only "active" citizens influence dispositional contagion (like in models of epidemics);
- Formulate all new features using simple formulas that are consistent with some applicable theory or have some empirical basis.

5.4.2 *Emergence*

The emergent properties of interest are the long-term behavior of the solutions (stability, intermittent bursts of rebellion, or permanent unrest), the patterns (distributions of size, duration and waiting time between successive events) of bursts of rebellion, and also the time variation of the number (or %) of "jailed" citizens, RD (if `RD?` = *true*) and legitimacy (if `LF?` = *true*).

Although the model is based on simple agents and rules, the multiplicity of parameters and mechanisms leads to complex behavior. Some parameters are expected to have associated tipping points. Also, the occurrence of large peaks of unrest is expected to be associated with some specific combinations of parameter values.

5.4.3 *Adaptation*

Agents have no adaptation capabilities.

[3]This speeds up the calculations because the numbers of "active," "jailed," and "quiet" agents are used more than once.

5.4.4 *Objectives*

In the present ABM the agents' objectives (goals) are encoded in a very simple way in their action rules. In the case of "activist" citizens, the additional goal of approaching concentrations of "active" citizens is encoded in their move rule. Since agents do not have cognitive capabilities or adaptive behavior, they do not rank decision alternatives according to some utility or fitness function.

5.4.5 *Learning*

Agents have no learning capabilities.

5.4.6 *Prediction*

In this model, "citizen" agents predict the net risk of turning rebellious based on the estimated arrest probability. However, they do not have predictive ability for decision making, in the sense of agents with cognitive or learning capabilities.

5.4.7 *Sensing*

Agents obtain information about the environment from two sources, the cells within their vision radius and, in the case of "citizens," the `group` and `infl` networks (if present). These sources correspond to distinct information spaces and variables, as well as different types of influence and percepts. Table 5.2 summarizes these aspects for the agent types and subtypes in the ABM.

5.4.8 *Interaction*

The interactions between the agents and the mechanisms related to them are summarized in Table 5.3.

Table 5.2 Agents' sensing: information space, type of influence, related state variables and percept (information used for decision/action) (Source: author)

Agent type	Information space	Type of influence	Related state variables	Percept
Citizen, "normal"	Cells within vision radius	Random citizens ("flock"/"mob")	v	# "active," "jailed," "quiet"; # "cop"
	`group` network	Group neighbors (two-way, strong ties)	`group-size`, `w-group`	"disposition" $G-N$
	`infl` network	"Activists", "influentials" (one-way, weak ties)	`infl-size`, `num-infl` `w-infl`	"disposition" $G-N$
Citizen, "activist"	Cells within vision radius	Random citizens ("flock"/"mob")	v	# "active", "jailed", "quiet"; # "cop"
	`group` network	Group neighbors (two-way, strong ties)	`group-size`, `w-group`	"disposition" $G-N$
Cop	Cells within vision radius	–	v'	"active" citizens

Table 5.3 Interactions between agents and related mechanisms (Source: author)

	"Citizens"	"Cops"
"Citizens"	• fear threshold lowering (via P_a) • massive fear loss (via P_a) • dispositional contagion (via $\sum(G-N)$) due to network influences	–
"Cops"	• deterrence/fear threshold rising (via P_a) • imprisonment (via imposing a `jail-term`)	–

5.4.9 *Stochasticity*

Stochasticity is essential to the working of the ABM. Many processes are modeled using pseudo-random variables, both in the initialization (`setup`) and in the model cycle (`go`).

In the initialization, pseudo-random variables are used to: (*i*) set up the initial positions of the agents; (*ii*) set up the "citizens" attributes, namely hardship, value (if `RD?` = *true*), and risk-aversion; (*iii*) build the `group` network, by randomly selecting the nodes of the group cliques and setting undirected links between them; and (*iv*) build the complete `infl` network, by randomly selecting the "activists" (hubs of directed star networks) and their "audience" among "citizen" agents.

In the model cycle, pseudo-random variables are used to: (*i*) define the agents' activation order; (*ii*) select the target cell in the agents' move rule; and (*iii*) select the "active" citizen to be arrested and the jail term in "cops" rule C.

The particular distributions of hardship, value, risk-aversion, and jail term play a key role in the generation of complex solutions. When exact reproducibility of the results is required, a random seed must be generated and stored in the output files. To reproduce a particular simulation, the user must set the random seed prior to running `setup`.

5.4.10 Collectives

The collectives represented in the ABM are the `group` and `infl` networks.

5.4.11 Observation

The observation data for the ABM are of two types: international indicators on legitimacy, human rights, economic decline, etc. for model parametrization; and databases of conflict events for analyzing and interpreting the solutions. In this work the FSI [70], FWI [28] and "All the Ginis" [49] indicators were used, as well as the SCAD database of conflict events [71], as described in Chap. 4.

5.4.12 Initialization

Model initialization is done in the `setup` procedure, as described in Sect. 5.3. Tables 5.4 and 5.1 describe the input parameters and the agent attributes, respectively.

5.4.13 Input Data

The model does not require any input from external data files. Parameter sweeping can be done using NetLogo's `BehaviorSpace` tool, but a simple R script based on the RNetLogo package [73] was written for that purpose, as described below.

5.4.14 Submodels

There are three important submodels: estimated arrest probability (i.e., the form of P_a used in the ABM), relative deprivation (for expressing the hardship as a function of RD), and (endogenous) legitimacy feedback.

Table 5.4 Description of the input parameters for the ABM of large scale conflict against a central authority (Source: author)

Parameter name	Description	Default value	Range
`world-width`	Width of model space	40	–
`world-height`	Height of model space	40	–
`initial-cop-density`	"cop" density (# cops/# cells)	4%	0%–100%
`initial-citizen-density`	"citizen" density (# citizens/# cells)	70%	0%–100%
`threshold`	Threshold (T)	0.1	0–1
`rho-c`	Critical C/A in P_a	0.5	0–5
`government-legitimacy`	Initial (or constant) legitimacy (L_0)	0.82	0–1
`max-jail-term`	Maximum jail term (J_{max})	30	0–400
`vision-a`	Vision radius for "citizen" agents (v)	7	0–30
`vision-c`	Vision radius for "cop" agents (v')	7	0–30
`RD?`	*true* if $G = f(RD)$, *false* otherwise	*false*	*true,false*
`RD-unif?`	*true* for `value` $\sim \mathscr{U}(0, 1)$, *false* otherwise	*false*	*true,false*
`gini`	Gini index for Pareto-distributed `value`	0.40	0–1
`gamma`	Exponent for power law RD (γ)	0.20	0.05 – 1.00
`GROUP?`	*true* if "group" network is present, *false* otherwise	*false*	*true,false*
`INFL?`	*true* if "infl" network is present, *false* otherwise	*false*	*true,false*
`group-size`	Group size	5	0–10
`infl-size`	Size of influence network associated with each "activist" (% population)	5%	0%–100%
`num-infl`	Number of "activists"	0	0–10
`w-group`	Weight for "group" dispositional contagion (if `GROUP?` = *true*)	1.0	0–10
`w-infl`	Weight for "infl" dispositional contagion (if `INFL?` = *true*)	1.0	0–10
`LF?`	*true* for legitimacy feedback, *false* otherwise	*false*	*true,false*
`LF-agents?`	*true* for heterogeneous legitimacy, *false* otherwise	*false*	*true,false*
`L-memory`	Number of time steps for averaging legitimacy (if `LF?` = *true*)	5	1–10
`k`	Arrest constant in P_a	2.3	2.3

The estimated arrest probability for "citizen" agents is computed using the following expression:

$$P_a(\rho) = \begin{cases} 0 & \rho < \rho_c \\ 1 - \exp(-k\,\rho) & \rho \geq \rho_c \\ 1 & \rho = +\infty \quad (A_v = 0) \end{cases} \tag{5.1}$$

where $k = 2.3$, $\rho = C_v/A_v$, and C_v and A_v are the numbers of "cops" and "active" citizens within the citizen's vision radius, respectively. Equation (5.1) is a generalization of one of the expressions considered by Fonoberova et al. [26], and includes equation (3.5) as a special case ($\rho_c = 0$). The parameter ρ_c can be used to model the mechanisms of "massive fear loss."

The RD submodel is activated by setting the input variable `RD?` to `true`. In this case, the hardship for each "citizen" is computed as follows. First, the "citizen" agent computes its "expectation" as the median of the `value` attribute of the other "citizens" within its vision radius. Then, it computes a "relative deprivation" as a function of the difference between its "expectation" and its own `value`:

$$\texttt{RD} = \max(\text{median}(\texttt{value})_v - \texttt{value}, 0) \tag{5.2}$$

where median(`value`)$_v$ is the "expectation". If the agent's `value` is lower than its "expectation," the `RD` is proportional to the difference between "expectation" and `value`, otherwise it is zero. The hardship H is then set equal to `RD`.

To compute the RD using (5.2) it is necessary to specify a distribution of `value`. In the ABM, this can be specified in two different ways, according to the Boolean input variable `RD-unif?`. For $\texttt{RD} - \texttt{unif?} = \textit{true}$, $\texttt{value} \sim \mathcal{U}(0, 1)$. For $\texttt{RD} - \texttt{unif?} = \textit{false}$, a Pareto Type I distribution is used [53]:

$$F_V(x) = \begin{cases} 1 - \left(\frac{x_m}{x}\right)^{\alpha} & x \geq x_m \\ 0 & x < x_m \end{cases} \tag{5.3}$$

where $\alpha > 0$ is the shape parameter and $x_m > 0$ is the scale parameter.[4] The Pareto-distributed value for "citizen" agents is generated using the formula

$$\texttt{value} = \frac{x_m}{U^{1/\alpha}} \tag{5.4}$$

where $U \sim \mathcal{U}(0, 1)$. In the code, $x_m = 2^{-(\alpha+1)/\alpha}$,[5] and the shape parameter is computed in terms of the input parameter `gini` using the equation

[4]The scale parameter x_m is the minimum possible value/level of welfare, e.g. the minimum national income. The shape parameter determines the inequality of value. It is necessary that $\alpha > 1$ for finite mean value and variance of the distribution defined by equation (5.3).

[5]This value of x_m is chosen so that the median of the distribution of `value` is 1/2, which is equal to the mean value of $H \sim \mathcal{U}(0, 1)$.

$$\alpha = \frac{\texttt{gini} + 1}{\texttt{gini}} \tag{5.5}$$

Equation (5.2) is a simplified model of economic deprivation, based on the hypothesis that deprivation is proportional to the difference between expectation and value.

In the case political RD, the values at stake (democracy, political participation, individual liberties) are different, and so is commitment and the emotional factor of deprivation. Both theory (e.g., [9, 37, 66]) and the SCAD exploration of conflict events in Sect. 4.1 favor the hypothesis that people are more sensitive to the gap between expectation and value in the case political deprivation than in the case of economic deprivation. This can be implemented in a simple way by using the following modified form of equation (5.2)

$$\texttt{RD} = \{\max(\text{median}(\texttt{value})_v - \texttt{value}, 0)\}^{\gamma} \tag{5.6}$$

where $\gamma > 0$ is another input parameter. $\gamma = 1$ corresponds to "economic RD," or to a "neutral" response to the gap between expectation and deprivation. If $0 < \gamma < 1$ the "citizens" become more sensitive (less tolerant) to this gap, as occurs with political issues.[6] Equation (5.6) is a straightforward implementation of proposition I.4 in Ted Gurr's frustration-aggression theory on the psychological factors of civil violence (Table 2.1, page 13).

The legitimacy feedback submodel was inspired in the theoretical framework of Gilley [33, 34] (Sect. 2.5). The perceived legitimacy L_p is expressed in terms of "views of legality" (L_{leg}), "views of justification"(L_{just}), and "acts of consent" (L_{cons}) as follows:

$$L_{\text{leg}} = 1 \tag{5.7}$$

$$L_{\text{just}} = \frac{n_{\text{quiet}}}{N_{\text{citizens}}} \tag{5.8}$$

$$L_{\text{cons}} = L_{\text{just}} \tag{5.9}$$

where N_{citizens} is the population size and n_{quiet} is the total number of "citizens" in the "quiet" state. Thus, it is assumed that the government is fully legal but justification and consent are expressed by the proportion of "citizens" that are not showing opposition. The legitimacy is computed using the "weighted average" formula proposed by Gilley [33, 34]:

$$L = L_0 \cdot \left(\frac{1}{4} \cdot (L_{\text{leg}} + L_{\text{cons}}) + \frac{1}{2} L_{\text{just}} \right) \tag{5.10}$$

[6] In the present work social RD is not considered, because ethnic, religious or other conflicts related to identity are not modeled.

$$= L_0 \cdot \left(\frac{1}{4} + \frac{3}{4} \cdot \frac{n_{\text{quiet}}}{N_{\text{citizens}}} \right) \tag{5.11}$$

where L_0 is the value of the `government-legitimacy` input parameter. For heterogeneous legitimacy perception (`LF-agents?` = *true*), n_{quiet} and N_{citizens} are replaced by the numbers of visible "citizens" in the "quiet" state and visible "citizens," respectively.

5.5 R Scripts for Pre- and Post-Processing

The model can be run interactively from the NetLogo GUI for quick visualization of tentative simulations, or off-line for conducting systematic experiments involving parameter sweeping, using the RNetLogo package [72, 73]. Two R-scripts were written, one for setting up and running a set of simulations with varying parameter values and storing the results in a `.RData` file, and another for post-processing the results. The pre-processing script does the following operations:

- Start NetLogo and load the model with default values of the input parameters (Table 5.4, page 60);
- Read the names of the input parameter(s) to be changed from a vector, and the values for each parameter from a list of numeric vectors (one vector of values for each parameter);
- Initialize a global list for storing the parameters and output for each run;
- Sweep over all parameter values using nested loops (one per parameter), and:
  - Set and store a new random seed[7];
  - Set the parameter values for the new run;
  - Run the model in an inner loop for a user-specified number of cycles;
  - Store a global list with (*i*) the random seed and full set of parameter values, and (*ii*) an output data frame with the time history of the proportion of "quiet", "active," and "jailed citizens," plus the (medians) of the legitimacy, expectation and deprivation, for each simulation.
- Save the run information (global list) in a `.RData` file for post-processing;
- Close the NetLogo instance.

The post-processing script reads a `.RData` file generated by the pre-processor, loops over all elements in the global list, and retrieves the values of the input parameters and the output data frame for each element. The retrieved information can then be used for further processing.

[7]This allows the exact reproducibility of the simulations, if required at a later time.

Chapter 6
Model Exploration and Computer Experiments

This chapter contains a description of a set of simulation experiments for exploration of the ABM proposed in the present work. In view of the large number of input parameters (density of "citizen" and "cop" agents, vision radii, legitimacy, threshold, ρ_c, γ, distributions of H and R, etc.) it was not feasible to perform an exhaustive exploration. Therefore, the simulation experiments were devised to illustrate the model's generative capacity, and to highlight the influence of the newly introduced mechanisms on the complexity of the solutions.

The first experiment, described in Sect. 6.1, illustrates the influence of the critical "cop"-to-"active" ratio ρ_c in the risk perception model on the size, duration, and recurrence of rebellion peaks. The relationship between ρ_c, the occurrence of cascades and the maximum possible peak size was demonstrated analytically and then studied via computer simulations. It was shown that the value of ρ_c has a strong impact on the stability of the system and has associated tipping points.

The second experiment, described in Sect. 6.2, was performed to show that the maximum jail term $J_{\max}$ strongly influences the interval between successive outbursts, but has a much smaller influence on the peak size and duration of the simulated events. No tipping points associated with $J_{\max}$ were found, for the setup conditions considered in the experiment.

Section 6.3 contains a description of simulations using the RD submodel (equation (5.6), page 62) to describe the hardship for different sensitivity to value (γ), in a scenario of low legitimacy and high level of repression. It was shown that the model can produce solutions with three different regimes (calm, punctuated equilibrium, and permanent turmoil) for low values of legitimacy. It was also found that the solutions' behavior is strongly dependent on the parameter γ, which has an associated tipping point for the conditions considered in the experiment.

Section 6.4 shows the results of computer experiments in which RD-dependent grievance is combined with legitimacy feedback, leading to complicated trajectories of the appropriate dependent variables (% of "active" and "jailed" agents, and medians of deprivation and legitimacy) in the corresponding phase spaces (resembling

C.M. Lemos, *Agent-Based Modeling of Social Conflict*, SpringerBriefs in Complexity, https://doi.org/10.1007/978-3-319-67050-8_6

those found in the study of chaotic systems), and to intermittent regime (long periods of calm, alternating with periods of turmoil) in an otherwise stable scenario of high legitimacy and low level of repression.

Finally, Sect. 6.5 contains a description of simulations including network influence effects, for the two different types of networks implemented in the model ("group" and "influentials"). It was found that network influence effects introduce instability, and that the degree of connectivity has a larger impact than the influence weight on the magnitude of the simulated events.

The results obtained in some of the experiments were discussed using the patterns of conflict events in the African countries affected by the AS obtained in Chap. 4 as background reference.

6.1 Risk Perception and the Estimated Arrest Probability

In this section, the influence of the parameter ρ_c in the estimated arrest probability function $P_a(\rho)$ defined by equation (5.1) will be analyzed in two stages. First it will be shown that, if network influence effects are not considered, the value of ρ_c in the definition of $P_a(\rho)$ determines the maximum possible size of the rebellion peaks. Large simulated peaks of turmoil only occur if the form of $P_a(\rho)$ leads to a drop of the risk perception for the whole population (herein called "massive fear loss") when the ratio between the number of "cops" and "active" citizens' falls below a critical value. This analysis also suggested a "rule of thumb" involving the legitimacy L, the threshold T and the estimated arrest probability for the model to produce solutions with intermittent peaks of rebellion. Then, computer experiments with varying ρ_c were done to confirm the findings of the analytical study. The mathematical analysis and the computer experiments were performed using the conditions in Run 2 of [21, 24] as reference case. Therefore, the hardship and risk aversion were set using uniform distributions and RD, legitimacy feedback and network effects were not considered. The numbers of "citizens" and "cops" were set $N_{\text{citizen}} = 1120$ and $C = 64$ respectively, legitimacy was set to $L = 0.82$, and the arrest constant was set $k = 2.3$.

6.1.1 Analytical Study

If network influence effects are not considered, the citizens' decision Rule A (page 53) can be written in the form

$$G > T + R \cdot P_a(\rho) \tag{6.1}$$

and expresses a balance between conflicting drives: motivation for rebelling on the left-hand side, and inhibition (fear of being arrested) on the right-hand side. The right-hand side of equation (6.1) can be interpreted as a "variable fear threshold" which depends on individual factors (R) and collective behavior ($P_a(\rho)$). Action is elicited once this variable threshold drops below the drive for rebelling (represented by the grievance G). Since $P_a(\rho)$ decreases with $\rho = (C_v/A_v)$, transitions from "quiet" to "active" result from lowering of the "variable fear threshold" due to other "citizen" agents turning rebellious.

In Epstein's ABM intermittent peaks of rebellion result from "citizen" agents turning "active" faster than "cops" can arrest them (by a cascade effect). The method for analyzing the possibility of occurrence of large cascades consists of considering the probability density function (pdf) of $G - N$, for different forms of $P_a(\rho)$ and given distributions of H and R. This can be imagined as a "1/2 Epstein ABM" without "cops" and arrests, to describe the "production" stage only, and corresponds to the method introduced by Granovetter for analyzing cascades in his threshold model of collective behavior [35].

The pdf for the grievance and risk perception in Epstein's original model [21, 24] are $G(L) \sim \mathscr{U}(0, 1 - L)$ and $N(\rho) \sim \mathscr{U}(0, P_a(\rho))$, respectively. The pdf and distribution of the random variable $X = G - N$ are [51]:

$$u_X(x, L, \rho) = \int_{-\infty}^{+\infty} g_G(z, L) \cdot n_N(z - x, P_a(\rho))\, dz \tag{6.2}$$

and

$$U_X(x, L, \rho) = \int_{-\infty}^{x} u_X(x', L, \rho)\, dx' \tag{6.3}$$

respectively, where g_G and n_N are the pdf of G and N. The number of "active" citizens is

$$A(L, \rho, T) = N_{\text{citizen}} \cdot (1 - U_X(L, \rho, T)) \tag{6.4}$$

and the corresponding proportion is $A(L, \rho, T)/N_{\text{citizen}}$.

Figures 6.1 and 6.2 show graphs of $u_X(x, L, \rho)$ for $L = 0.82$ and $\rho_c = 0$ and $\rho_c = 1$, respectively, considering four different values of ρ. For uniform distributions of H and R, and estimated arrest probability given by equation (5.1), the expressions for the pdf and distribution of $G - N$ are very involved. Therefore, the calculations were done using *Mathematica* [79].

When $\rho_c = 0$, the pdf has the same trapezoidal variation for all values of ρ (Fig. 6.1). The pdf is flatter and the probability density is mainly concentrated on negative values of x for high levels of deterrence ("cop"-to-"active" ratio), and becomes progressively taller and concentrated on positive values of x for low levels of deterrence. When $\rho_c = 1$, the pdf has a trapezoidal variation similar to that of the previous case when the level of deterrence is high, but below the critical value $\rho = 1$ the probability density functions change into a pulse function in $[0, 1 - L]$ (Fig. 6.2).

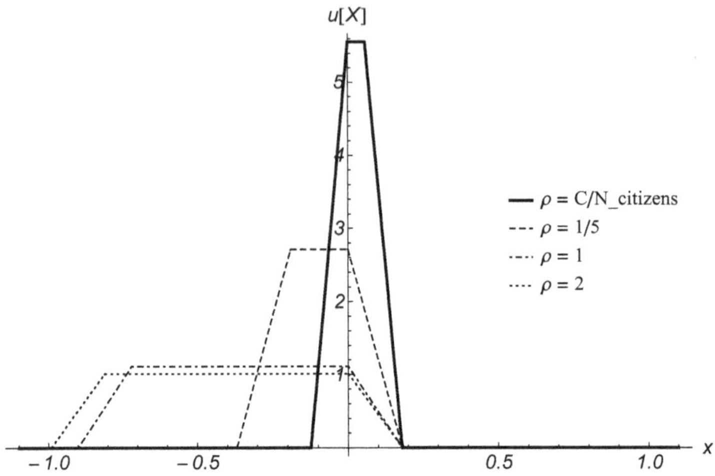

Fig. 6.1 Probability density functions $u_X(x, L, \rho)$ for $L = 0.82$, $\rho_c = 0$ and four different values of ρ (Source: author)

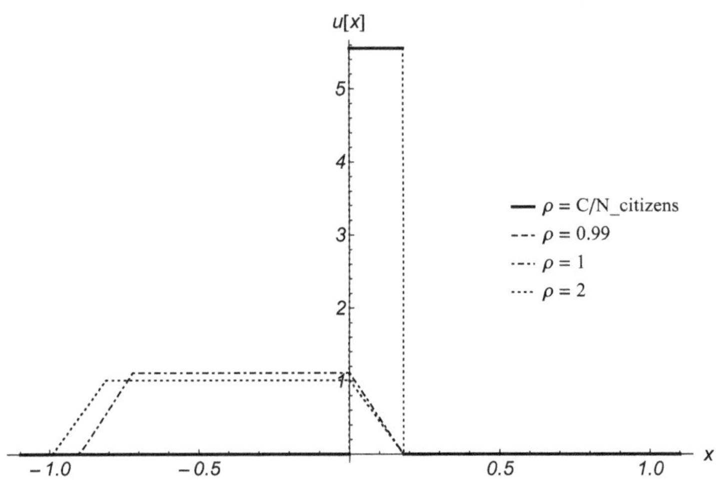

Fig. 6.2 Probability density functions $u_X(x, L, \rho)$ for $L = 0.82$, $\rho_c = 1$ and four different values of ρ (Source: author)

Figure 6.3 shows the variation of the proportion of "active" citizens with ρ for $T = 0.05$ and $T = 0.1$, and for $\rho_c = 0$ and $\rho_c = 1$. When $\rho \geq 1$ the proportions are small and nearly identical for both values of ρ_c. But for $\rho_c = 1$ the proportions suddenly increase to the maximum possible value (which depends on the value of T) for $\rho < 1$.

These results can also be confirmed by means of the fixed point iteration $A_{n+1} = N_{\text{citizen}} \cdot (1 - U_X(L, C/A_n, T))$, which can be written in terms of ρ as $\rho_{n+1} = C/A_n$

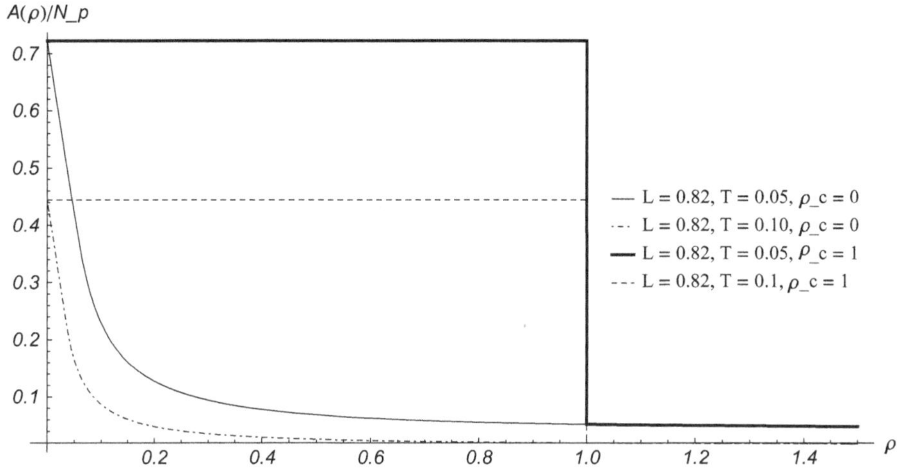

Fig. 6.3 Proportion of "active" citizens as a function of ρ for $L = 0.82$ and two values of the threshold ($T = 0.05$ and $T = 0.1$), with $G(L) \sim \mathscr{U}(0, 1 - L)$ and $N(\rho) \sim \mathscr{U}(0, P_a(\rho))$ (Source: author)

(Fig. 6.4). For $\rho_c = 0$ there exists a fixed point ρ_f which corresponds to a relatively high value of C/A and low proportion of "active" citizens. For $\rho_c = 1$ one or two fixed points may exist, but there is always one fixed point for $\rho \ll 1$ which corresponds to a high proportion of "active" citizens.

In summary, for uniform distributions of H and R, the parameter ρ_c in $P_a(\rho)$ given by equation (5.1) has an important influence on the maximum possible size of the rebellion peaks. Inclusion of this parameter in the expression for the estimated arrest probability provides a mathematical representation of a mechanism of "massive fear loss," by which the risk perception (or "fear of retribution") disappears for the whole population when the level of deterrence falls below a critical value.

The analytical results relating the form of the estimated arrest probability and ρ_c to the maximum possible magnitude of the rebellion peaks can be checked via simulation, by setting up a "1/2 Epstein ABM" with only "citizen" agents, endowed with Rules M and A, and with C_v replaced by the mean value of the number of "cops" that "citizens" would see (total number of "cops" multiplied by the number of cells within the vision radius v, and divided by the number of cells in the model space). Running this "1/2 Epstein ABM" for a few cycles, it is possible to observe the evolution of the simulated distribution of $G - N$ and compare the % of "active" citizens with analytical predictions.

Figures 6.5, 6.6, 6.7, 6.8 illustrate one such experiment. A "1/2 Epstein ABM" was built using NetLogo [77], with a torus space with 120×120 cells, $N_{\text{citizen}} = 10080$ "citizens" and $C = 576$ fictitious "cops." The legitimacy and threshold were

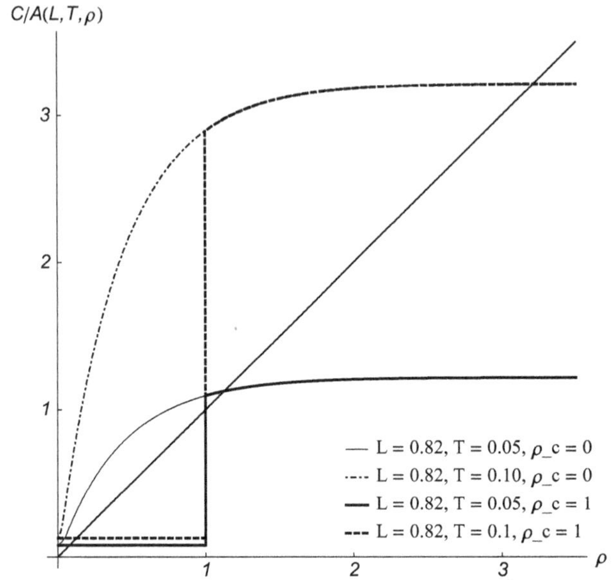

Fig. 6.4 Graphical representations of $C/A(L, T, \rho)$ vs ρ for $L = 0.82$, $T = 0.05$ and $T = 0.1$, for $\rho_c = 0$ and $\rho_c = 1$ (Source: author)

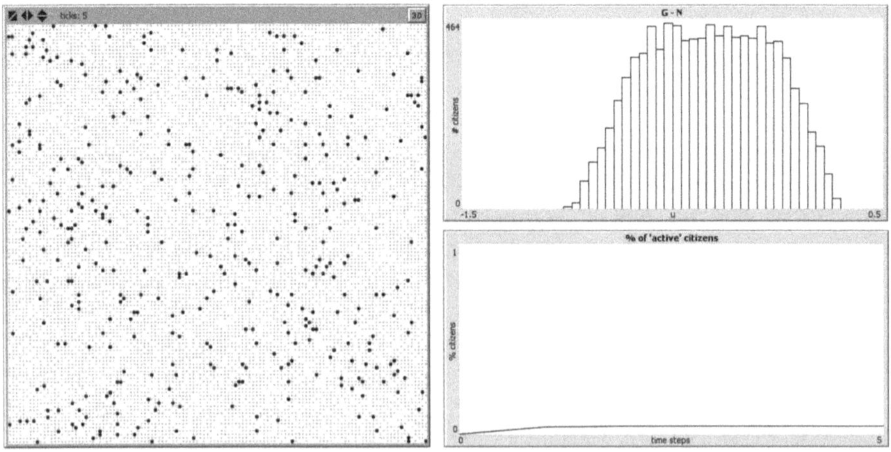

Fig. 6.5 Model space (*left*), simulated histogram of $G - N$ (*top right*), and proportion of "active" citizens (*bottom right*) for the "1/2 Epstein ABM" with $\rho_c = 0$, after five time cycles. "Quiet" and "active" citizens are represented by *gray and black dots*, respectively (Source: author)

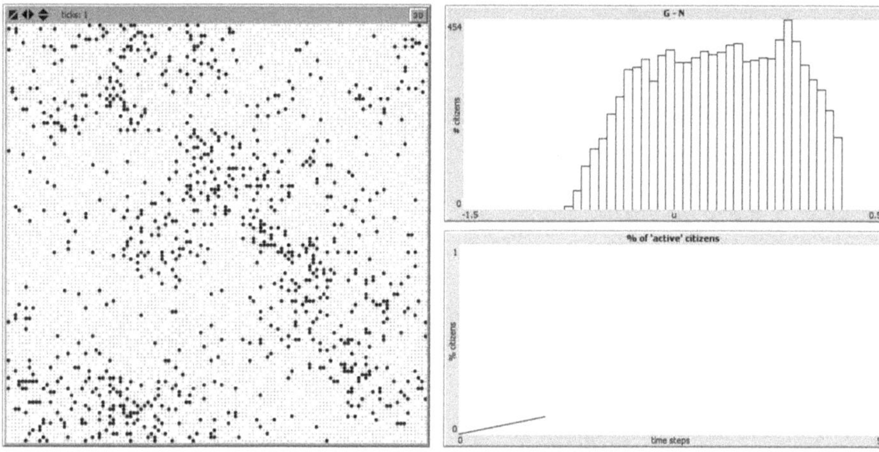

Fig. 6.6 Model space (*left*), simulated histogram of $G - N$ (*top right*), and proportion of "active" citizens (*bottom right*) for the "1/2 Epstein ABM" with $\rho_c = 1.0$, first cycle. "Quiet" and "active" citizens are represented by *gray and black dots*, respectively (Source: author)

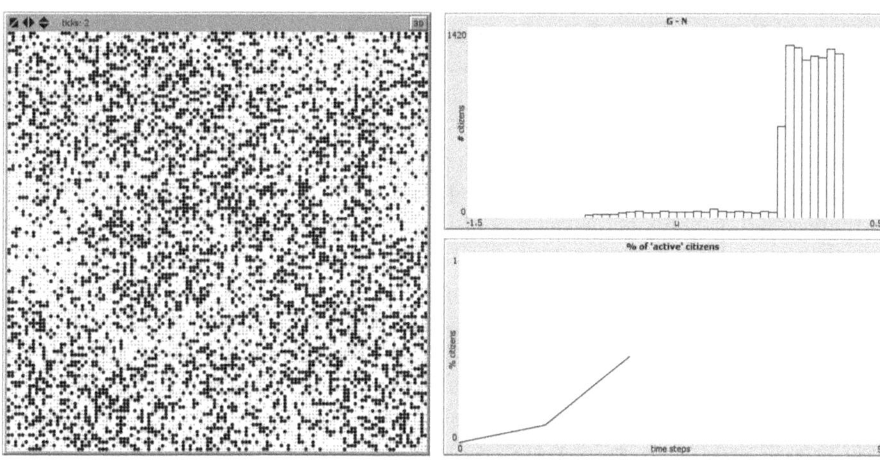

Fig. 6.7 Model space (*left*), simulated histogram of $G - N$ (*top right*), and proportion of "active" citizens (*bottom right*) for the "1/2 Epstein ABM" with $\rho_c = 1.0$, second cycle. "Quiet" and "active" citizens are represented by *gray and black dots*, respectively (Source: author)

set $L = 0.70$ and $T = 0.15$, respectively, and the vision radius was set $v = 7$.[1] Two simulations were performed, with $\rho_c = 0$ and $\rho_c = 1$. Both simulations started with all "citizens" in the "quiet" state.

[1]This setup of grid size, N_{citizen} and C leads to the same "citizen" and "cop" densities as in the reference conditions in Run 2 of [21, 24], but with an almost tenfold increase of the number of agents. This was done to yield clearer histograms and better estimates of the maximum % of "active" citizens. The legitimacy was set lower than in the reference conditions so that the initial histograms of $G - N$ show a more pronounced trapezoidal shape.

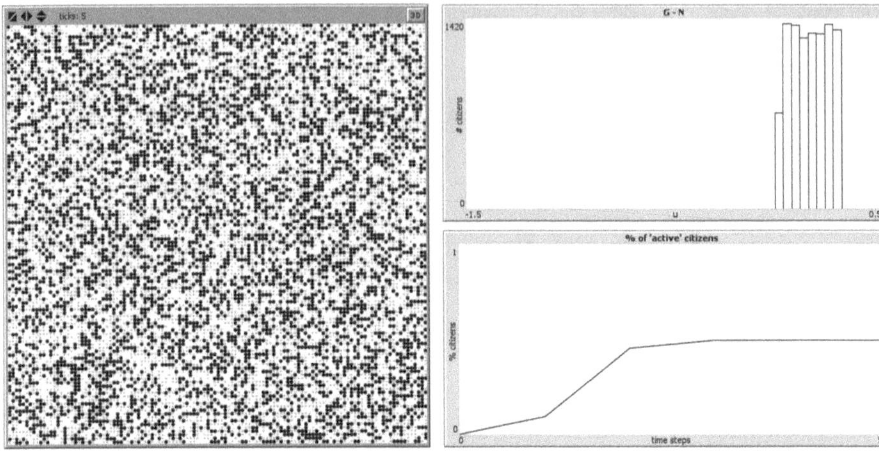

Fig. 6.8 Model space (*left*), simulated histogram of $G - N$ (*top right*) and proportion of "active" citizens for the "1/2 Epstein ABM" with $\rho_c = 1.0$, fifth cycle. "Quiet" and "active" citizens are represented by *gray and black dots*, respectively (Source: author)

Figure 6.5 shows the model space, simulated histogram of $G-N$ and time history of the % of "active" citizens after five cycles (steady state solution) of a simulation with $\rho_c = 0$. The simulated histogram of $G - N$ has a trapezoidal shape similar to that of the pdf of $G - N$ in Fig. 6.1. For the conditions of this simulation, large cascades of "citizens" turning rebellious are not possible. Indeed, the computed % of "active" citizens after five cycles was 4.08%, for a theoretical value of 3.88%.

Figures 6.6, 6.7, 6.8 show the model space, simulated histogram of $G - N$ and time history of the % of "active" citizens for cycles one, two, and five of a simulation with $\rho_c = 1$. After one cycle (Fig. 6.6), the simulated histogram of $G-N$ still shows a trapezoidal shape, like that in the previous simulation. However, at the end of the second cycle (Fig. 6.7) there was a massive cascade of "citizens" turning "active." The simulated histogram shows the bars for high frequencies shifted to the right (towards positive values of $G - N$), forming the rectangular pulse shape of the pdf of $G - N$ shown in Fig. 6.2 for $\rho < \rho_c$. Figure 6.8 shows the steady state solution after five cycles. In this case, the computed % of "active" citizens was 49.95%, for a theoretical value of 50.00%.

The snapshots of the evolution of the simulation with $\rho_c = 1$ show why this parameter can be thought as a model representation of a mechanism of "massive fear loss:" once the "cop"-to-"active" ratio drops below a critical value (ρ_c), the risk perception suddenly vanishes for the whole society, and grievance ceases to be moderated by fear. This leads to a massive cascade of "citizens" rebelling.

* * *

The analytical study and simulations described above clarify the importance of L and $P_a(\rho)$, and why some forms of the latter lead to solutions with large rebellion peaks and other forms do not. This was noted but not explained by Wilensky [78] and Fonoberova et al. [26].

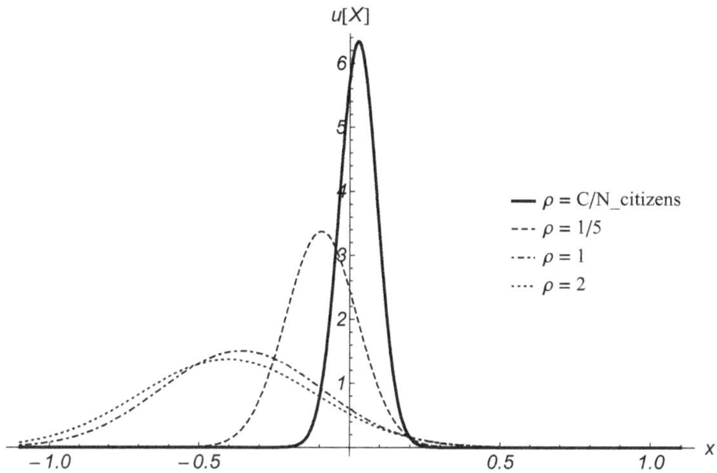

Fig. 6.9 Probability density functions $u_X(x, L, \rho)$ for normally distributed H and R, $L = 0.82$, $\rho_c = 0$ and four different values of ρ (Source: author)

This finding can be further confirmed by repeating the analysis for normal distributions for G and N, with the same mean values and variances as the uniform distributions, in the citizens' action rule. In this case $G(L) \sim \mathcal{N}((1-L)/2, (1-L)/\sqrt{12})$ and $N(\rho) \sim \mathcal{N}(P_a(\rho)/2, P_a(\rho)/\sqrt{12})$ respectively,[2] and $G - N$ is normally distributed with $\mu(G-N) = (1 - L - P_a(\rho))/2$ and $\text{var}(G-N) = \left((1-L)^2 + P_a^2(\rho)\right)/12$ [51].

Figures 6.9 and 6.10 show the pdf $u_X(x, L, \rho)$ when G and N are normally distributed, with means and variances as stated above, for $\rho_c = 0$ and $\rho_c = 1$, respectively. The qualitative behavior is the same as for uniform distributions, i.e. for $\rho_c = 0$ the pdf is progressively narrower and shifted to the right for increasing ρ, whereas for $\rho_c = 1$ all pdf are reduced to the pdf of G for $\rho < \rho_c$.

Figures 6.11 and 6.12 show the maximum proportion of "active" citizens and the variation of $C/A(L, \rho, T)$ as a function of ρ for $L = 0.82$, threshold $T = 0.05$ and $T = 0.1$, and $\rho_c = 0$ and $\rho_c = 1$, for normally distributed G and N. The qualitative behavior of the expected solutions is similar to the previous case, in the sense that $\rho < \rho_c$ leads to sudden vanishing of the risk perception for the whole population and to solutions with larger peaks of rebellion than would occur if $\rho_c = 0$. However, these peaks are smaller when G and N are normally distributed than when they follow uniform distributions, for the same values of L, T, $\mu(G)$ and $\mu(N)$.

[2]This implies that the supports of g_G and n_N will be $]-\infty, +\infty[$, so that the hardship H and risk aversion R may take values outside the interval [0, 1], which violates the premises on the range of these variables for the case of uniform distributions. However, the probability of H or R falling outside [0, 1] is small.

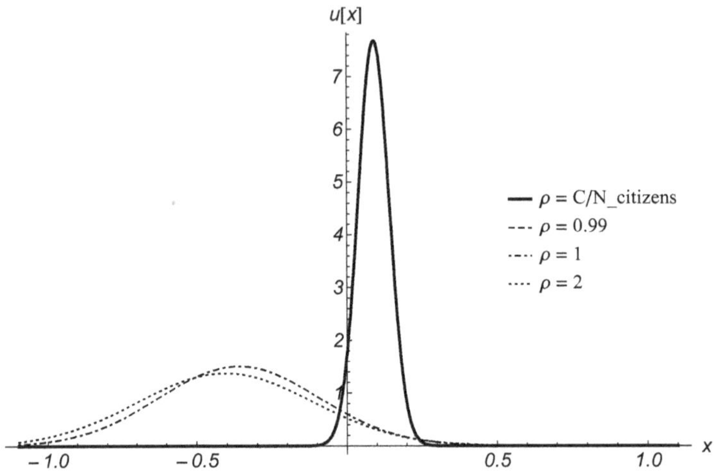

Fig. 6.10 Probability density functions $u_X(x, L, \rho)$ for normally distributed H and R, $L = 0.82$, $\rho_c = 1$ and four different values of ρ (Source: author)

The analysis above also suggests the following qualitative condition ("rule of thumb") between L, $P_a(\rho)$ and T for a significant proportion "citizens" turning "active":

$$T \leq \frac{1}{2}(1 - L - P_a(\rho)) + \sqrt{\frac{(1-L)^2 + P_a^2(\rho)}{12}} \tag{6.5}$$

It is important to consider the theoretical justifications for the use of a threshold decision rule with an "aggression" and a "risk inhibition" drive and for selecting the form of the estimated arrest probability function. Ethology provides a very general and low-level conceptual foundation for using a threshold rule like Epstein's Rule A. According to Lorenz [46], aggression (harmful action towards members of one's own species) has a fundamental evolutionary value and behavioral patterns in animals result from a tension between conflicting drives, namely aggression (attack) and fear (escape). The threshold for initiating an attack is lowered by situational factors. This tension-threshold model, combined with the mechanism of redirection of aggression to avoid excessive intra-specific damage, explains cyclic patterns of animal fights and confrontation rituals. The tension-threshold model is also consistent with the key elements of micro-situational theories of violence (e.g., [11, 12, 76]).

There are also theoretical arguments in favor of modeling $P_a(\rho)$ with functions for which the perceived risk drops to zero below a critical value of ρ. The occurrence of large protests and uprisings after long periods of calm in real processes also supports the conjecture that such forms are more plausible than a simple exponential variation.

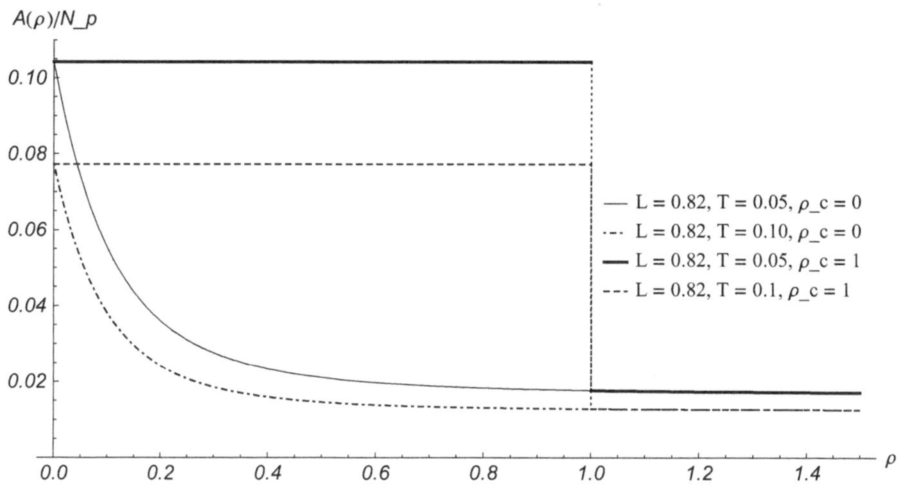

Fig. 6.11 Proportion of "active" citizens as a function of ρ for $L = 0.82$ and two values of the threshold ($T = 0.05$ and $T = 0.1$), with $g_G \sim \mathcal{N}((1-L)/2, (1-L)/\sqrt{12})$ and $n_N \sim \mathcal{N}(P_a(\rho)/2, P_a(\rho)/\sqrt{12})$ (Source: author)

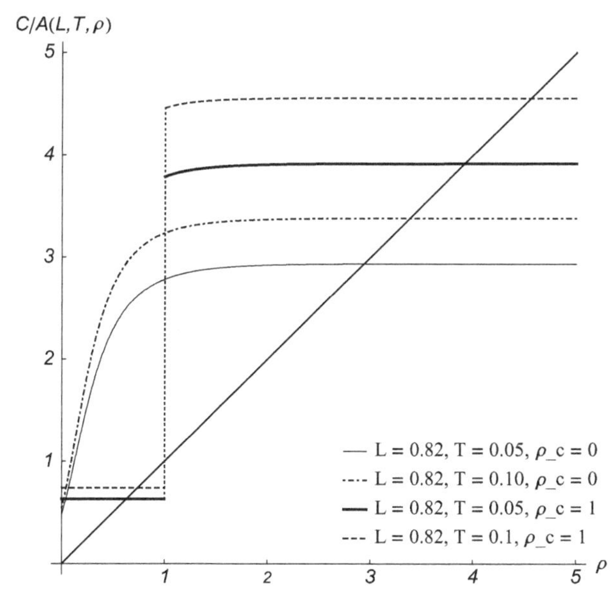

Fig. 6.12 Graphical representations of $C/A(L, T, \rho)$ vs ρ for $L = 0.82$, $T = 0.05$ and $T = 0.1$, for $\rho_c = 0$ and $\rho_c = 1$, with $u_X(x, L, \rho)$ for normally distributed H and R (Source: author)

6.1.2 Computer Experiments

This set of computer simulations was performed to investigate the influence of the parameter ρ_c on the behavior of the solutions. For $\rho_c = 0$ the model reduces to Epstein's ABM. The value $\rho_c = 0.25$ was adopted by Fonoberova et al. [26], and $\rho_c = 1.0$ approximately corresponds to the formula for the estimated arrest probability proposed by Wilensky [78]. In this set of experiments all extensions to Epstein's model (RD-dependent grievance, legitimacy feedback and network influences) were turned off.

One hundred runs were performed for values of ρ_c between 0 and 2, in steps of $\Delta\rho_c = 0.025$. The remaining input parameters were set to their default values in Table 5.4 (page 60), which are representative of the Run 2 simulation reported in [24]. This default setup corresponds to an artificial society of 1120 "citizens" and 64 "cops" (with total density 74%), relatively high `government-legitimacy` (0.82) and low `threshold` (0.1). Each simulation was run for 5,000 cycles, to obtain a number of simulated activity peaks allowing the calculation of meaningful statistics of size, waiting time and duration of rebellion events. The parameters `peak.threshold` and `diff.threshold` used for detecting the peak size, interval and duration of significant rebellion bursts in the post-processing stage were set to their default values (0.1% and 0.05%, respectively).

Figure 6.13 shows the mean values of the mean and maximum size of the simulated conflict events in % of the number of "citizen" agents (a), and the mean

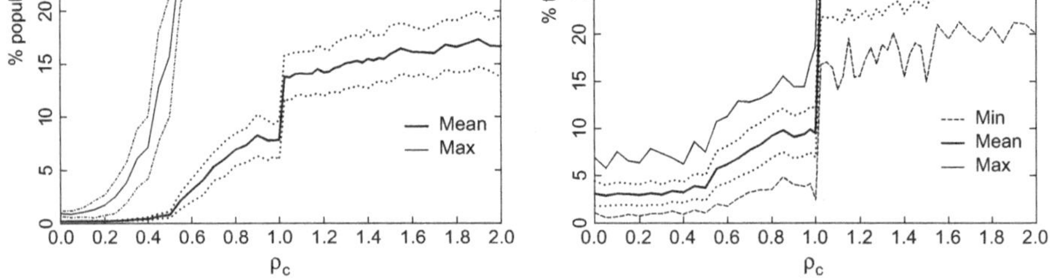

Fig. 6.13 Mean values of the mean and maximum size of rebellion peaks (% of "active" citizens) (**a**), and mean values of the minimum, mean and maximum % of time with rebellion (**b**) for the ρ_c variation experiment, computed using 100 output files. The spacing between *dotted and solid lines* of mean peak size (**a**) and % of time with rebellion (**b**), and the *dot-dashed lines* above and below the maximum peak size in (**a**), represent one standard deviation of the corresponding variables (Source: author)

values of minimum, mean and maximum % of time with conflict events (b), for the ρ_c variation experiment, computed using 100 output files. It is observed that for $\rho_c < 0.5$ the mean peak size remains below 1%. The maximum peak size is $\sim$ 1% for $\rho_c = 0$ and starts to increase at $\rho_c \sim 0.2$, reaching $\sim$ 5% at $\rho_c \sim 0.3$. At $\rho_c = 0.5$ the mean peak size is still about 1% but starts to increase at a larger rate. The maximum peak size increases at a faster rate for $\rho_c > 0.4$ reaching $\sim$ 15% for $\rho_c = 0.5$. For $\rho_c = 1.0$ the mean peak size jumps suddenly and the maximum peak size reaches $\sim$ 30%.

The variation of % of time with rebellion events in Fig. 6.13b also shows that for $\rho_c < 0.5$ the mean value is approximately constant, but for $\rho_c > 0.5$ it starts to increase. At $\rho_c \sim 1$ there is a sudden jump of the mean value of the % of time with rebellion events from $\sim$ 8% to $\sim$ 25%, which then increases slowly for $\rho_c > 1$. This suggests that for the densities of "citizen" and "cop" agents, legitimacy, threshold and distributions of H and R, there is a tipping point associated with ρ_c for $\rho_c \sim 1$ and possibly another for $\rho_c \sim 0.5$. It is clear that ρ_c strongly influences the qualitative and quantitative properties of the solutions, particularly the size of rebellion events, confirming the conclusions of the analytical study.

Figure 6.14 shows the mean values of the mean and maximum event duration (a) and interval between successive events (b), in time steps, for the ρ_c variation experiment, computed using 100 output files. There is a correspondence between Figs. 6.13a and 6.14a, as the variations of mean and maximum values of the respective statistics are qualitatively similar. This was to be expected, since larger peaks also take longer for "cops" to suppress them.

Figures 6.13b and 6.14b show that larger % of time with turmoil corresponds to shorter intervals between successive events. Figure 6.14b also shows another interesting fact, namely that the mean interval between successive events is approximately constant for $\rho_c < 1$ and $\rho_c > 1$ with a discontinuous drop at $\rho_c = 1$. This is due to the fact that while the size of rebellion peaks is strongly dependent on ρ_c, the interval between successive events is strongly influenced by the maximum jail term $J_{\max}$, as shown in Sect. 6.2 below.

Recalling Fig. 4.7, it can be observed that for demonstrations before the AS the typical and maximum % of time with such events in the countries analyzed are $\sim$ 1% and $\sim$ 14%, respectively, and that after the beginning of the AS these figures increased to $\sim$ 5% and $\sim$ 26%. Thus, $\rho_c \leq 0.2$ and $0.2 \leq \rho_c \leq 0.5$ are plausible ranges of values for representing the % of time with social unrest in (relatively) stable and (clearly) unstable situations, respectively, like those before and after the beginning of the AS.

From the analysis in Sect. 4.1, it is not possible to obtain upper bounds for the maximum estimated % of the population involved in huge demonstrations and riots, for comparison with the values obtained in the simulations. Nevertheless, it seems reasonable to assume that peaks as large as 30% are unrealistic. This suggests that the condition $\rho_c < 1.0$ is required for meaningful correspondence between simulated and real events in large scale conflict processes.

Another way of showing how ρ_c influences the behavior of the solutions is to observe the simulated time histories of % of "active" citizens for different

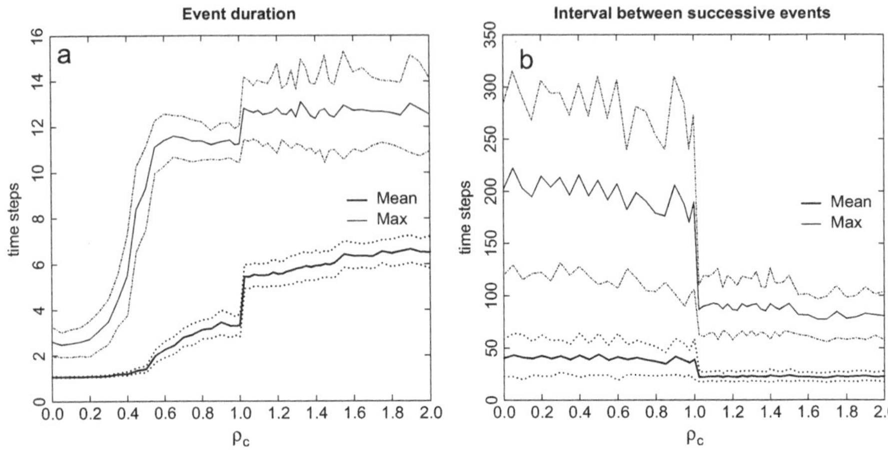

Fig. 6.14 Mean values of the mean and maximum event duration (**a**) and interval between successive events (**b**) in time steps, for the ρ_c variation experiment, computed using 100 output files. The width between the *solid and dotted lines* above and below the mean and maximum values represents one standard deviation of the corresponding variables (Source: author)

values of this parameter. Figure 6.15 shows the first five hundred cycles for four representative cases with $\rho_c \in [0, 1]$. The simulation with $\rho_c = 0$ (top) corresponds to the conditions of the reference simulation (Run 2 in [24]). The simulation for $\rho_c = 0.25$ (value adopted by Fonoberova et al. [26] for studying urban crime) has qualitative properties similar to those found in, e.g., demonstrations in Morocco and Tunisia before the beginning of the AS, in terms of the amplitude of peak size. The simulation for $\rho_c = 0.5$ shows more frequent intermittent bursts of rebellion, as well as larger peaks and a wider range of peak sizes. This corresponds to a greater instability of the society, closer to the situation after the beginning of the AS. The simulation with $\rho_c = 1.0$ shows peaks larger than 25% which are probably unrealistic.

Figures 6.16 and 6.17 show box plot representations of the distributions of peak size and event duration, obtained by pooling the peak sizes and event durations of all events retrieved from the output files, for four selected values of ρ_c. The information in these figures is consistent with the previous conclusions.

The results of this computer experiment can be summarized as follows:

- The parameter ρ_c in equation (5.1) has a strong impact on the behavior of the solutions;
- The computer experiments suggest that, for the combination of parameters in a reference simulation (Run 2 in [24] and [21]), there are tipping points associated with ρ_c;
- The values 0.25 and 0.5 are useful for setting ρ_c in further explorations, for they lead to solutions with meaningful qualitative correspondence with conflict events in societies with moderate or strong conflict intensity (e.g., the conditions before and after the beginning of the AS), respectively. The choice $\rho_c = 1.0$ leads to peaks of unrealistic magnitude.

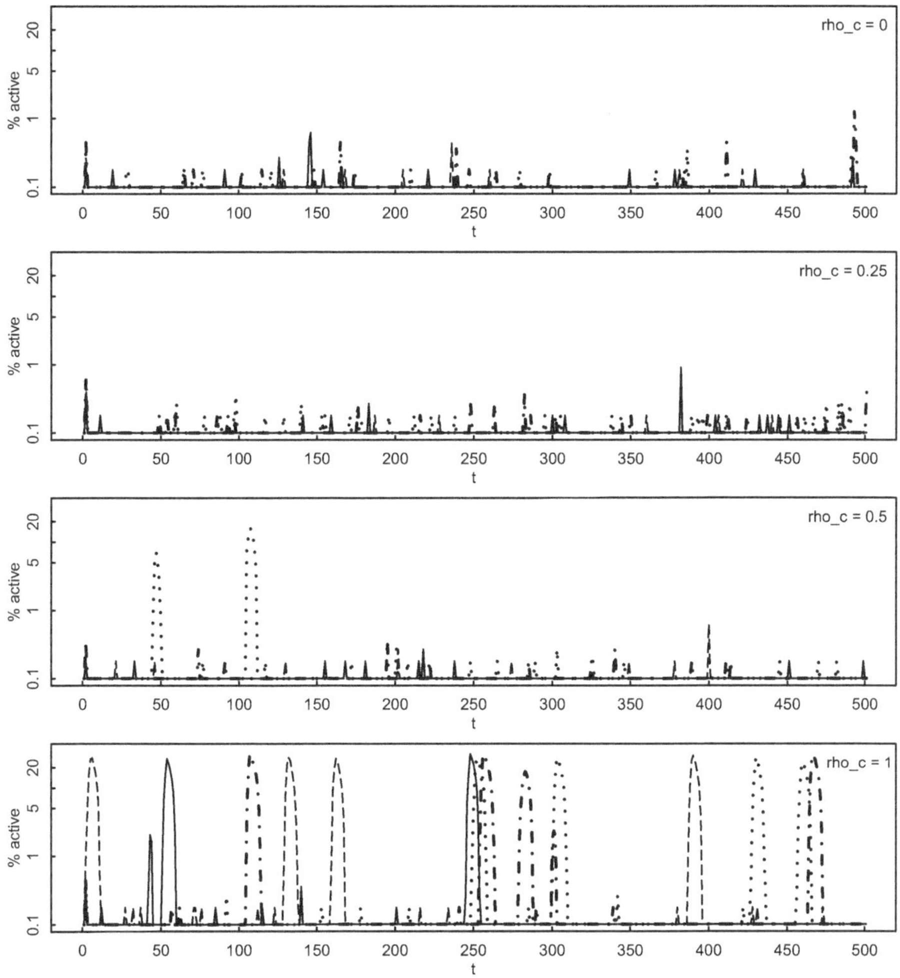

Fig. 6.15 Simulated time history of the proportion of "active" citizens for the first five hundred cycles of four different simulations of the $\rho_c = 0.0$, $\rho_c = 0.25$, $\rho_c = 0.50$ and $\rho_c = 1.00$ cases (Source: author)

6.2 The Influence of the Jail Term

In the previous experiment, it was shown that ρ_c has a strong impact on the size of rebellion peaks, but a smaller influence on the interval between successive events. The purpose of this experiment was to evaluate the influence of J_{max} on the behavior of the solutions, as done for ρ_c.

One hundred simulations were performed for values of J_{max} between 10 and 60, in steps of $\Delta J_{max} = 5$. The remaining input parameters were set to their default

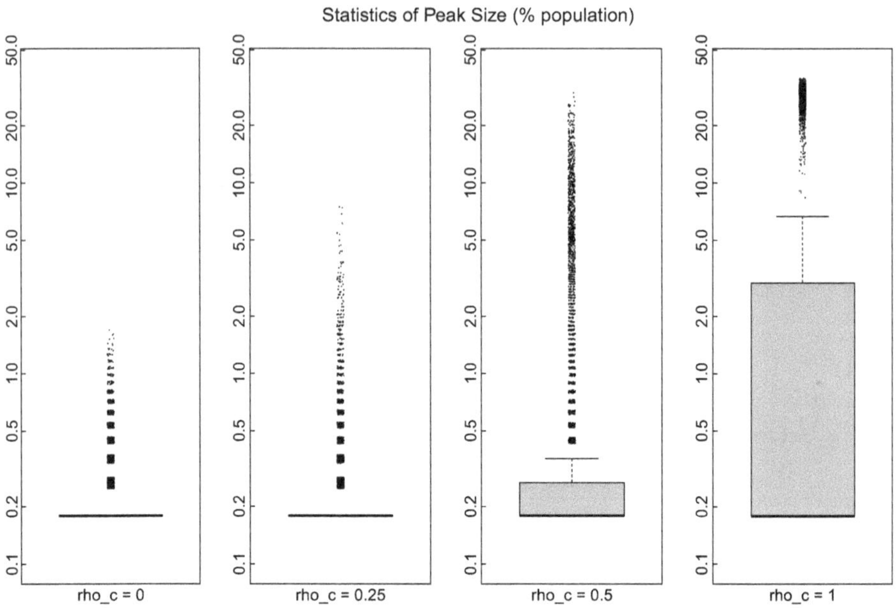

Fig. 6.16 Box plot summary representation of the distributions of rebellion peak size (% of "active" citizens) for $\rho_c = 0$, 0.25, 0.5, and 1, obtained by pooling the peak sizes retrieved from 100 output files for each case. Jitter was added to the coordinates of the outliers to help visualizing their distribution (Source: author)

values (Table 5.4, page 60), which correspond to the same general setup of the previous experiment (small artificial society, high `government-legitimacy` and low `threshold`). Each simulation was run for 5,000 cycles, and the parameters `peak.threshold` and `diff.threshold` used for detecting the peak size, interval and duration of significant rebellion bursts in the post-processing stage were set to 0.1% and 0.05%, respectively.

Figure 6.18 shows the mean values of the mean and maximum peak size of the simulated conflict events (a), and the mean values of the minimum, mean and maximum % of time with conflict events (b), for the J_{max} variation experiment, computed using 100 output files.

It can be observed that variations of J_{max} have a smaller influence on the mean peak size than variations of ρ_c, although the maximum peak size decreases between $J_{max} = 10$ and $J_{max} = 50$. This can be explained as follows: as J_{max} is increased, the release of "jailed" citizens is slower, and "cops" suppress new uprisings with a larger % of "citizens" still "jailed" from the previous event. The reasons for the non-monotonic variation of the maximum peak size for $J_{max} > 50$ are unclear.

The minimum, mean and maximum proportions of % of time with rebellion shown in Fig. 6.18b decrease monotonically with the variation of J_{max}, in a way that is consistent with the variations of event duration and interval between successive events. This is illustrated in Fig. 6.19, which shows the mean values of the mean and maximum event duration (a) and interval between successive events (b) in time steps, for the J_{max} variation experiment, computed using 100 files.

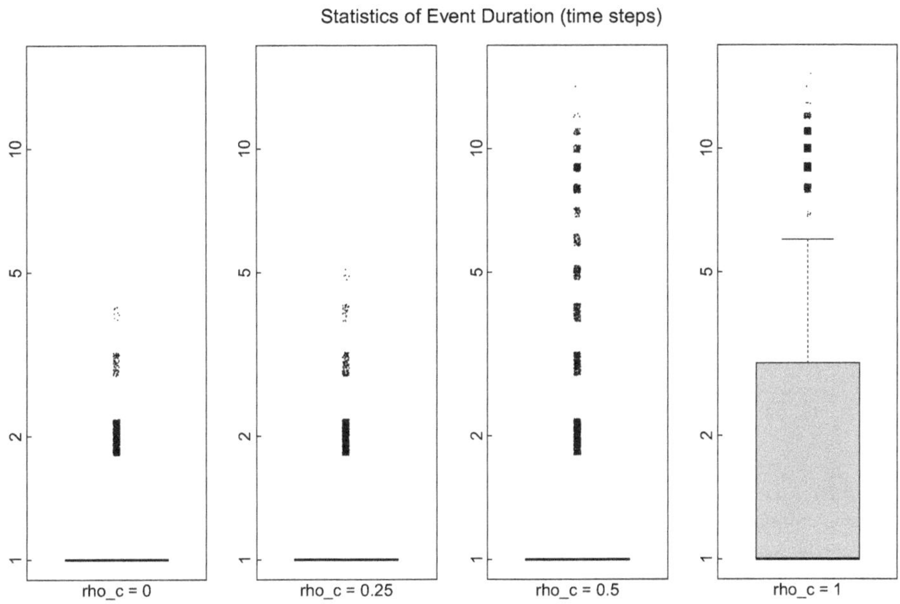

Fig. 6.17 Box plot summary representation of the distributions of rebellion burst duration (in time steps) for $\rho_c = 0, 0.25, 0.5$, and 1, obtained by pooling the event durations retrieved from 100 output files for each case. Jitter was added to the coordinates of the outliers to help illustrating their distribution (Source: author)

Figures 6.20, 6.21, 6.22 show box plot representations of the distributions of peak size, event duration and interval between successive events, obtained by pooling the peak sizes, event durations and event intervals retrieved from 100 output files, for four selected values of J_{max}. These figures provide confirmation that, for the conditions of this experiment, J_{max} has a stronger influence on the interval between successive events than on the peak size and event duration. In contrast with ρ_c, no tipping points associated with J_{max} were identified.

It is interesting to consider how to set J_{max} for the simulations to have meaningful correspondence with conflict events in real processes, like those analyzed in Sect. 4.1.4 for the case of the African AS countries. This is difficult for a number of reasons. For instance, the statistics of event interval are different among the countries and for the period considered (before and after the beginning of the AS). Also, it is not possible to define a precise correspondence between the time scale of one day real events as described in SCAD, and one time step in the ABM. Nevertheless, it is possible to envisage at least a plausible range for J_{max}, as follows.

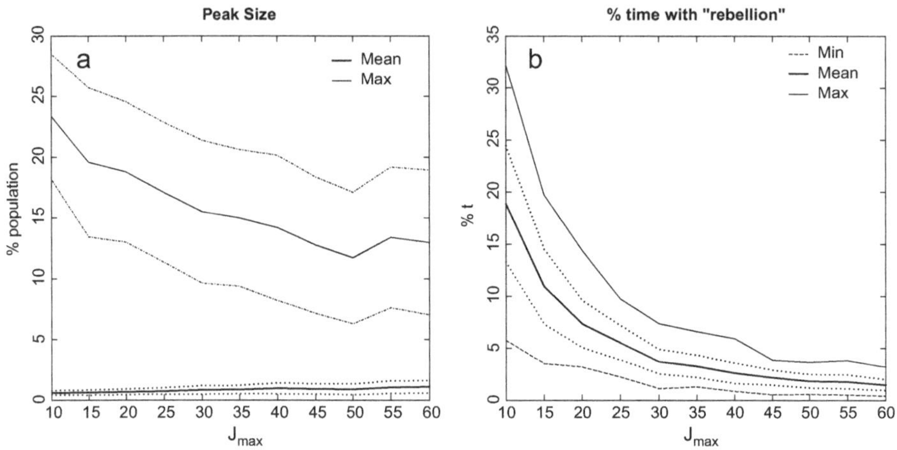

Fig. 6.18 Mean values of the mean and maximum size of rebellion peaks (% of "active" citizens) (**a**), and minimum, mean and maximum % of time with rebellion (**b**) for the J_{max} variation experiment, computed using 100 output files. The spacing between *dotted and solid lines* of mean peak size (**a**) and % of time with rebellion (**b**), and the *dot-dashed lines* above and below the maximum peak size in (**a**), represent one standard deviation of the corresponding variables (Source: author)

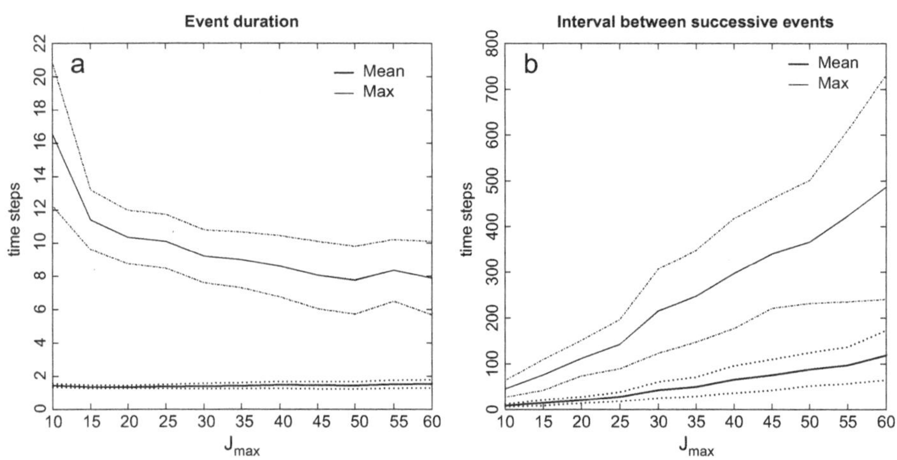

Fig. 6.19 Mean and maximum event duration (**a**) and interval between successive events (**b**) in time steps, for the J_{max} variation experiment, computed using 100 output files. The width between the *solid and dotted lines* above and below the mean and maximum values represents one standard deviation of the corresponding variables (Source: author)

Table 6.1 shows summary statistics of event interval for demonstrations in African AS countries after 2010-12-15[3] and the simulations of the J_{max} variation

[3]The period after the beginning of the AS was selected, because it is representative of manifest large-scale conflict and the variability among the countries is smaller.

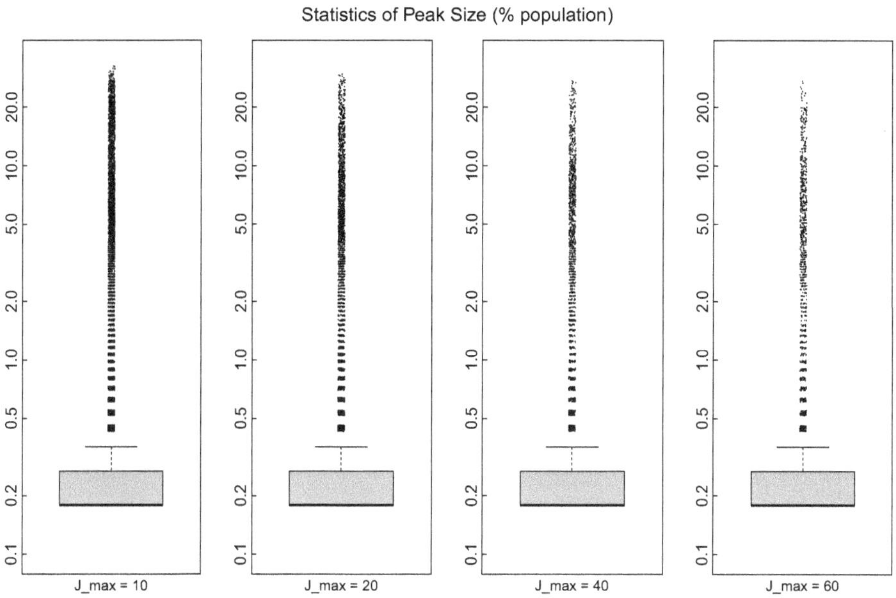

Fig. 6.20 Box plot summary representation of the distributions of rebellion peak size (% of "active" citizens) for $J_{max} = 10, 20, 40$, and 60, obtained by pooling the peak sizes retrieved from 100 output files for each case. Jitter was added to the coordinates of the outliers to help visualizing their distribution (Source: author)

experiment. Except for the cases of Egypt and Mauritania, the range for the ratio between event interval and duration is [10,40]. For the conditions of this experiment (size and density of the artificial society, vision radii, distributions of individual attributes, legitimacy, threshold, etc.) a plausible range for J_{max} is [20,40], which includes the default value $J_{max} = 30$ in the reference simulation (Epstein et al. [24], Run 2).

6.3 Relative Deprivation

The experiments in Sects. 6.1 and 6.2 were based on Run 2 in references [21, 24], in which $L = 0.82$ and $T = 0.1$. This combination of high legitimacy and low threshold, often used in studies based on Epstein's model (e.g., [26, 52]), can be considered typical of democratic regimes. In the context of the present work, it is important to evaluate the model's generative capacity for describing more realistic scenarios of governments facing generalized popular revolt. It is assumed that such scenarios are characterized by low legitimacy, sufficiently high deterrence capability (to prevent permanent turmoil), and high sensitivity to deprivation (e.g., political RD related to freedom and human rights). These conditions can be considered representative of authoritarian regimes.

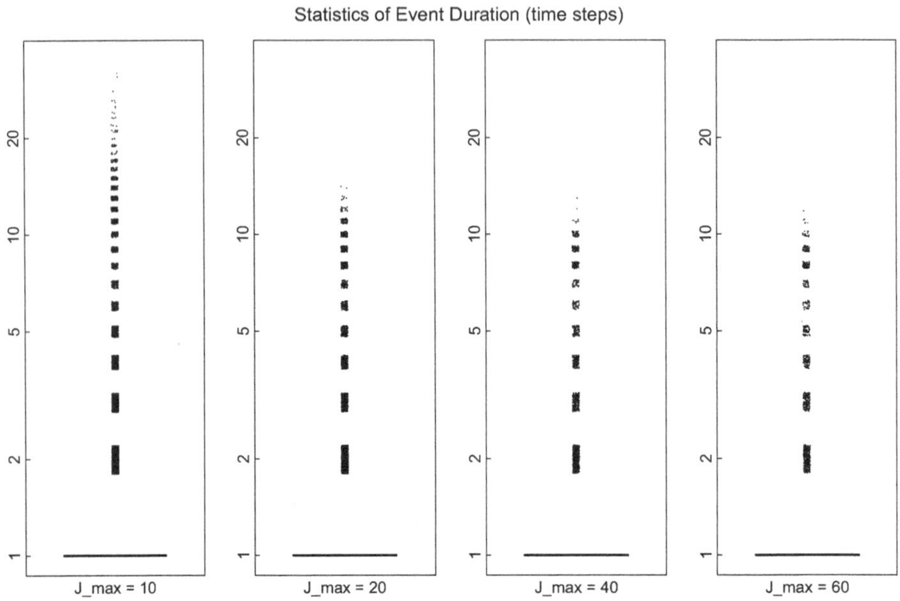

Fig. 6.21 Box plot summary representation of the distributions of rebellion burst duration (in time steps) for $J_{\max} = 10$, 20, 40, and 60, obtained by pooling the event durations retrieved from 100 output files for each case. Jitter was added to the coordinates of the outliers to help illustrating their distribution (Source: author)

The purpose of the experiment described in this section was to investigate the model's capability for generating solutions with stability, punctuated equilibrium (intermittent uprisings), and permanent instability (continuous rebellion) for low legitimacy, and to show how the solutions' behavior changes for varying deterrence capability and sensitivity to RD. This experiment is also relevant to confirm the validity of inequality (6.5) as a stability condition for both low and high values of legitimacy, and to evaluate the present model's capability for representing different forms of RD (political vs economic).

Therefore, the values of legitimacy and welfare inequality were set consistent with the indicators for the AS countries studied in Chap. 4 and the distribution of hardship was specified as a function of RD using equation (5.6). The legitimacy was set to 0.20 and the Gini index to its default value 0.40. For complex solutions with intermittent bursts of rebellion to occur, inequality (6.5) gives an estimate $T \sim 0.6$, which was the value used in the simulations. The maximum jail term was set to 40.

Since the size of the artificial society has a significant impact on the solutions' behavior [45], two sets of simulations were performed using different grid sizes, one representing a "small society" and another a "large society." The thresholds for minimum % of the population for a "significant" rebellion event, used in the post-processing stage, were 10^3 and 10^4 for the "small" and "large" society settings, respectively. Table 6.2 shows the input parameters for this experiment that differ from the default values shown in Table 5.4 (page 60).

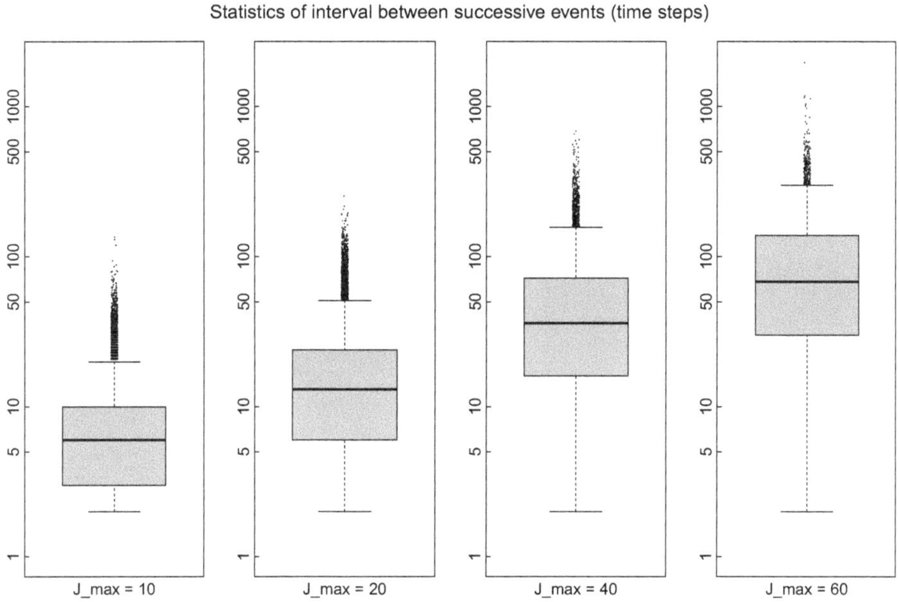

Fig. 6.22 Box plot summary representation of the distributions of intervals between successive events (in time steps) for $J_{max} = 10, 20, 40$, and 60, obtained by pooling the event intervals retrieved from 100 output files for each case. Jitter was added to the coordinates of the outliers to help illustrating their distribution (Source: author)

The sweeping parameters were the `initial-cop-density`, which is related to the deterrence capability of the central authority (e.g., [24]), and the γ exponent in equation (5.6), which is related to the sensitivity to deprivation (commitment, or emotional factor). The values of γ chosen for this experiment were all smaller than 1, corresponding to varying degrees of "political deprivation." The number of simulations for each pair of values of the sweeping parameters was 100 for the "small society" and 20 for the "large society" settings, respectively. Each simulation was run for 5,000 cycles.

Figures 6.23 and 6.24 show level plots of the mean values of the median of the % of time with rebellion for the simulations in this experiment, for the "small" and "large society" settings, respectively.[4] Both figures show that depending on the combination of the two sweeping parameters, there are three distinct regimes: permanent rebellion (for small `initial-cop-density` and `gamma`), complex solutions with intermittent peaks of rebellion, and solutions with permanent calm. It can also be observed that γ is a critically important parameter, for the solutions' behavior is more sensitive to variations of γ than to those of the "cop" density.

[4]The medians were computed for each of the output files available in the "small" and "large society" cases (100 and 20, respectively), and then the mean value was computed. This provided estimates for both the central tendency, which is of primary interest, and for the dispersion across different simulations of the same case.

Table 6.1 Summary of statistics of interval between successive large demonstrations in African "Arab Spring" countries after 2010-12-15, and rebellion bursts in simulations of the $J_{\max}$ variation experiment ($\rho_c = 0.5$), obtained by pooling the event intervals retrieved from 100 files (Source: author, based on model simulations and [71])

Large demonstrations	Min.	Median	Mean	Max.	N_{events}
Algeria	3.0	12.5	57.8	546	16
Egypt	0.0	3.0	5.2	49.0	200
Libya	0.0	18.5	32.2	259	32
Mali	2.0	36.0	74.2	377	9
Mauritania	65.0	180	199	367	5
Morocco	0.0	15.0	26.3	98.0	39
Sudan	1.0	21.0	35.4	131	26
Tunisia	0.0	12.0	21.4	243	51
Simulations, $J_{\max} =$					
10	2	6.00	8.25	135	65258
15	2	10.00	14.12	213	40597
20	2	14.00	20.13	253	27417
25	2	19.00	27.49	360	20379
30	2	28.00	42.82	539	13454
35	2	32.00	49.56	547	11675
40	2	42.00	65.69	685	9028
45	2	49.00	76.61	689	7626
50	2	59.00	88.50	952	6426
55	2	64.00	97.02	1373	5982
60	2	81.0	120.72	1967	4851

In the "small society" case (Fig. 6.23) variations of γ have a stronger impact on the solutions' behavior than variations of the "cop" density, but if the latter is sufficiently high rebellion is suppressed. The region in the parameter space for which the solutions have intermittent bursts of rebellion with realistic peak sizes has a staircase-like shape. For $\gamma \geq 0.150$, variations of the "cop" density have little influence on the % of time with rebellion. This suggests that there is a tipping point associated with γ, at $\gamma \sim 0.150$.

In the "large society" case, variations of γ have a stronger impact on the solutions' behavior than in the "small society" case: the region of instability is larger, and for small values of γ it would take a higher "cop" density than explored in the experiment for the solutions to be stable. For $\gamma \geq 0.150$, transition from intermittent rebellion to calm is abrupt and almost independent of the "cop" density. This suggests the existence of a tipping point associated with γ between 0.150 and 0.175 (Fig. 6.24).

Figures 6.25 and 6.26 illustrate the mean values of the mean and maximum % of time with rebellion in three-dimensional surface plots, for the "small" and "large society" cases, respectively. These figures show that the difference between mean and maximum values is not significant, in the sense that both statistics of

Table 6.2 Values of the input parameters and number of "citizen" and "cop" agents for the RD experiment, for the cases of "small society" and "large society" (Source: author)

	"Small society" experiment	"Large society" experiment
Parameter name	Value(s)	Value(s)
world-width	40	120
world-height	40	120
initial-cop-density	2.0% to 5.0% with step 0.25 %	2.0% to 5.0% with step 0.25%
initial-citizen-density	70%	70%
Number of "citizen" agents	1120	10080
Number of "cop" agents	32 to 80 with step 4	288 to 720 with step 36
threshold	0.60	0.60
government-legitimacy	0.20	0.20
max-jail-term	40	40
RD?	*true*	*true*
gamma	0.05 to 0.225 with step 0.025	0.05 to 0.225 with step 0.025

this variable (% of time with rebellion) can be used to determine the qualitative behavior of the solutions (particularly for the "large society" case). The difference between the "small" and "large society" cases is very clearly shown. Figure 6.26 again suggests the existence of a tipping point associated with γ between 0.150 and 0.175 for the "large society" case.

In short, the model predicts that if the cause of RD is sufficiently sensitive, it will be difficult for the central authority to avoid rebellion—governments must be very aware of the sensitivity of RD they impose on the population, otherwise they may face massive uprisings, unless their deterrence capability and repression level are very high.

It is interesting to ask: for which combinations of values of `initial-cop-density` and γ are the simulated % of time with rebellion consistent with the typical values for large demonstrations in the African countries after the beginning of the AS, analyzed in Chap. 4 (Sect. 4.1.4)?

For the "small society" case these combinations of values are represented in blue, dark, and light purple in Fig. 6.23. For the "large society case" these combinations are shown in the same colors, which roughly correspond to $3.00 \leq$ `Initial.cop.density` ≤ 5.00 and $\gamma = 0.15$. In both cases, increasing the level of "political RD" can be modeled by decreasing γ.

Figures 6.27 and 6.28 show level plots of the mean values of the median and maximum peak size (% of the population) of the rebellion bursts for the simulations in this experiment for the "small" and "large society" settings, respectively. The `NA` values represented in white, correspond to solutions with absolute calm or with a proportion of "active" citizens below the threshold. These figures show the same variation of the solutions' qualitative behavior with variations of `Initial.cop.density` and γ obtained by analyzing the % of time with

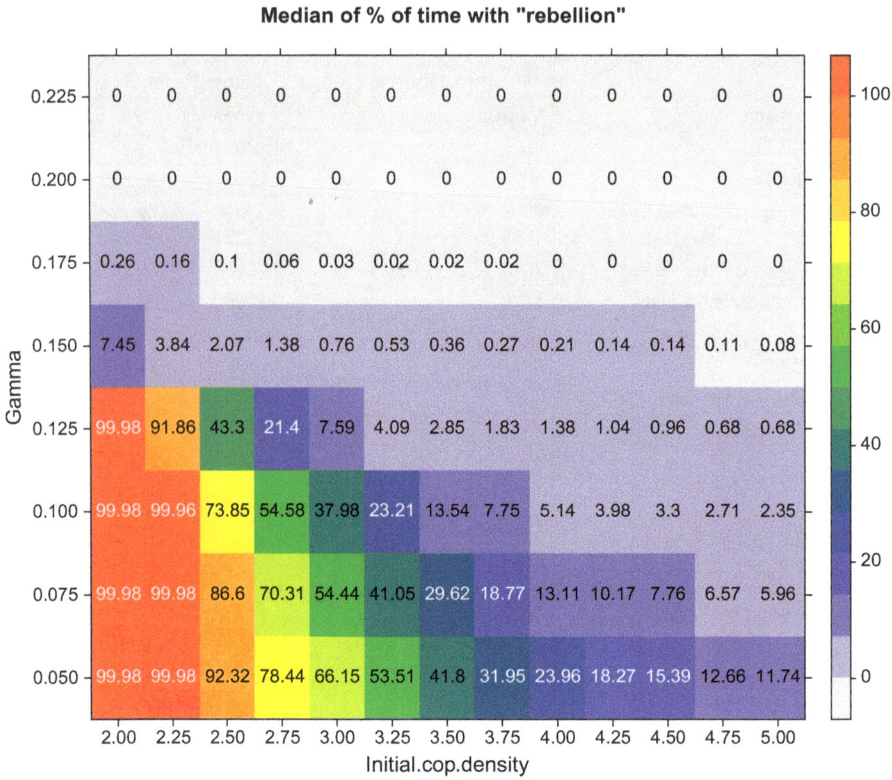

Fig. 6.23 Level plot of the mean value of the median of the % of time with rebellion for the RD experiment, for the "small society" case, computed from 100 simulations for each pair of values of the sweeping parameters (Source: author)

rebellion. However, the differences between the (averaged) median and (averaged) maximum peak size are larger than for the mean % of time with rebellion. This is further illustrated in Figs. 6.29 and 6.30, which show the relationship between the variations of the mean values of the median, mean and maximum peak sizes (% of population) obtained in the simulations, for the "small" and "large society" cases, respectively. The (averaged) median of peak size decreases more sharply than the (averaged) mean peak size for increasing `Initial.cop.density`, particularly in the "small society" case. Taking the analysis of large demonstrations in African countries affected by the AS as reference (Chap. 4, Sect. 4.1.4), $0.125 \leq \gamma \leq 0.150$ can be considered as the range for solutions to have plausible values for the maximum peak size, for the conditions of this experiment (citizens' density, vision radii of "citizen" and "cop" agents, ρ_c, etc.).

Figures 6.31 and 6.32 show level plots of the mean values of the median and maximum event duration (in time steps) for this experiment, for the "small" and "large society" settings, respectively. The `NA` values represented in white,

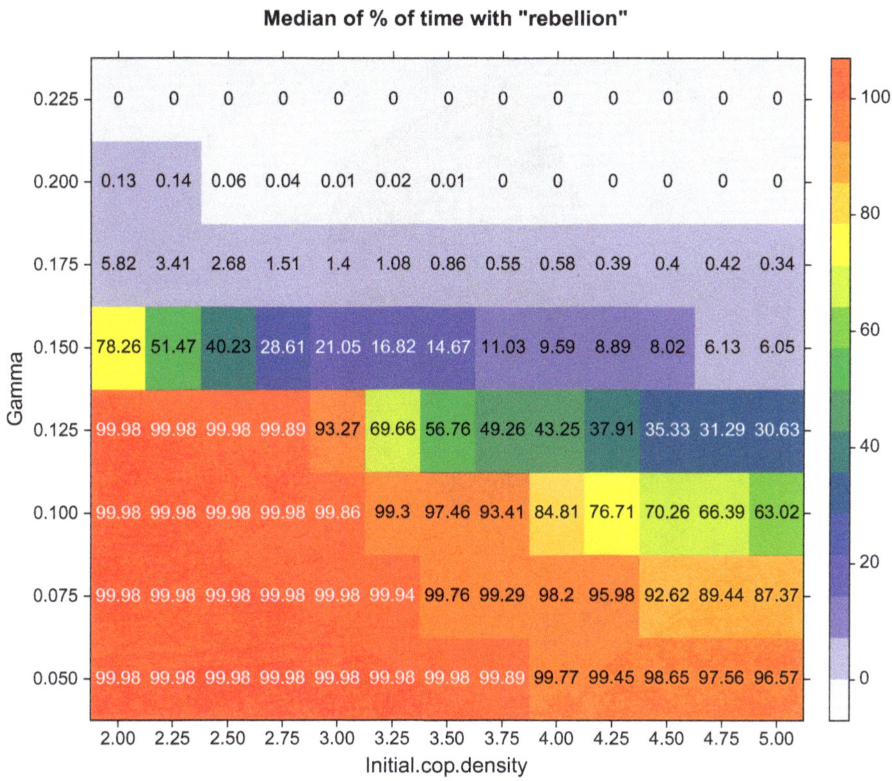

Fig. 6.24 Level plot of the mean value of the median of the % of time with rebellion for the RD experiment, for the "large society" case, computed from 20 simulations for each pair of values of the sweeping parameters (Source: author)

correspond to solutions with absolute calm or with a proportion of "active" citizens below the threshold. It is observed that the event duration is larger for the "large society" case, for all combinations of values of the sweeping parameters (it takes longer for "cops" to suppress peaks of rebellion in large societies). Overall, $\gamma \sim 0.15$ leads to event durations that have plausible correspondence with those in large demonstrations (or riots) analyzed in Chap. 4, Sect. 4.1.4.

The change of qualitative behavior associated with changes of sensitivity to RD can also be analyzed by examining the time history of the medians of expectation and deprivation (taken over all "citizens") for stable and unstable solutions. Figure 6.33 shows the first 500 cycles of the medians of expectation and deprivation in the artificial society for two different situations (pairs of values of the sweeping parameters). It can be observed that for $\gamma = 0.15$ the median of expectation exceeds that of deprivation, whereas for $\gamma = 0.10$ the reverse is true. The variations of the expectation and deprivation are also qualitatively different in the two simulations. In the $\gamma = 0.15$ case, the median of the expectation remains almost constant and the

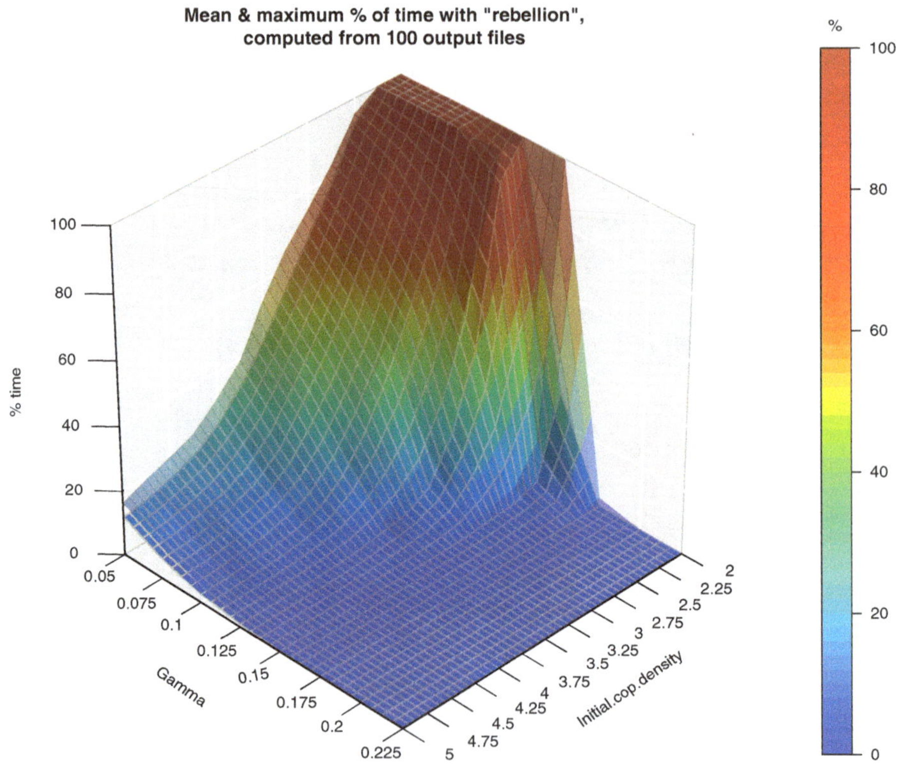

Fig. 6.25 Surface plots of the mean value of the mean (opaque) and maximum (semi-transparent) % of time with rebellion for the RD experiment for the "small society" case, computed from 100 simulations for each pair of values of the sweeping parameters (Source: author)

median of deprivation oscillates very rapidly, never surpassing 0.5. In the $\gamma = 0.10$ case, there are periods in which both medians oscillate, with that of deprivation always exceeding the one of expectation.

6.4 Legitimacy Feedback

In the previous sections of this chapter, it was shown that the risk perception (controlled by the ρ_c parameter) and the sensitivity to RD (associated with parameter γ) considerably influence the long-term behavior of the solutions (stability, punctuated equilibrium or permanent rebellion). Since legitimacy is a key variable, it is also important to study the influence of endogenous legitimacy feedback (i.e., how does the state of the system, described by the proportion of rebellious and/or jailed citizens, affects the legitimacy).

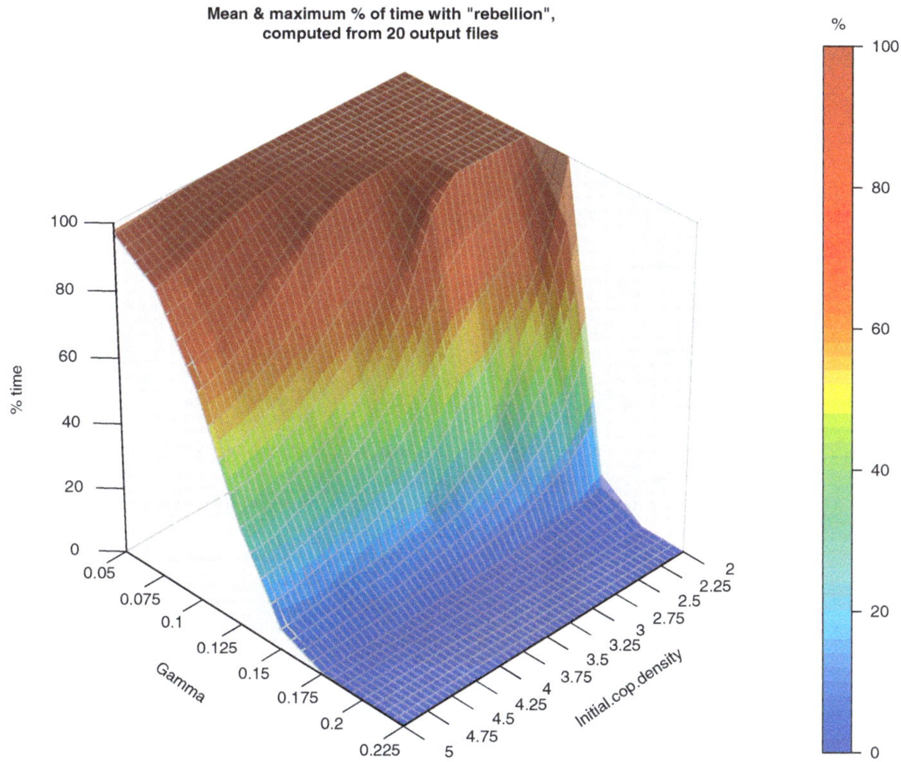

Fig. 6.26 Surface plots of the mean value of the mean (opaque) and maximum (semi-transparent) % of time with rebellion for the RD experiment for the "large society" case, computed from 20 simulations for each pair of values of the sweeping parameters (Source: author)

In this section, two computer experiments will be described. The first was performed to study whether or not legitimacy feedback changes the solutions' behavior in the scenario of low legitimacy and high level of repression considered in Sect. 6.3. The second was designed to show how legitimacy feedback leads to solutions with intermittent alternations of regime (long periods of calm followed by periods of unrest, with unpredictable transitions between them), in a scenario of high legitimacy and low level of repression.

6.4.1 Relative Deprivation with Legitimacy Feedback

The purpose of the experiment described below was to evaluate whether or not the introduction of endogenous legitimacy feedback changes the behavior of the solutions, for the conditions of the experiment described in Sect. 6.3. More specifically, the goal was to investigate the effects of varying levels of deterrence and

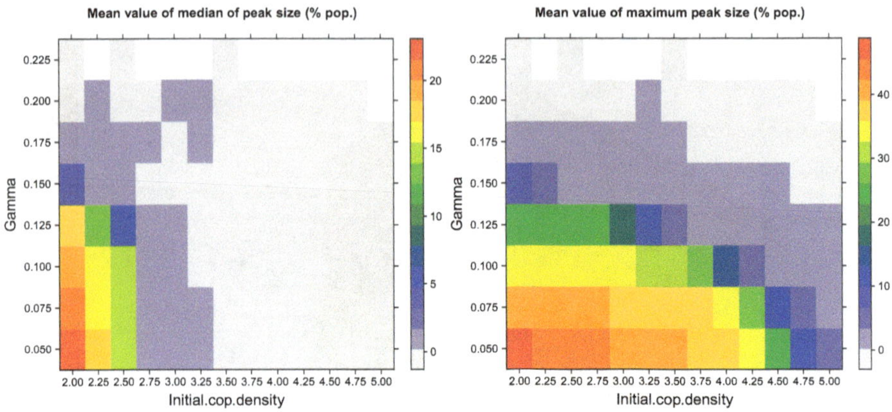

Fig. 6.27 Level plots of the mean values of the median and maximum peak size (% of the population) of rebellion bursts for the RD experiment, for the "small society" case, computed from 100 simulations for each pair of values of the sweeping parameters (Source: author)

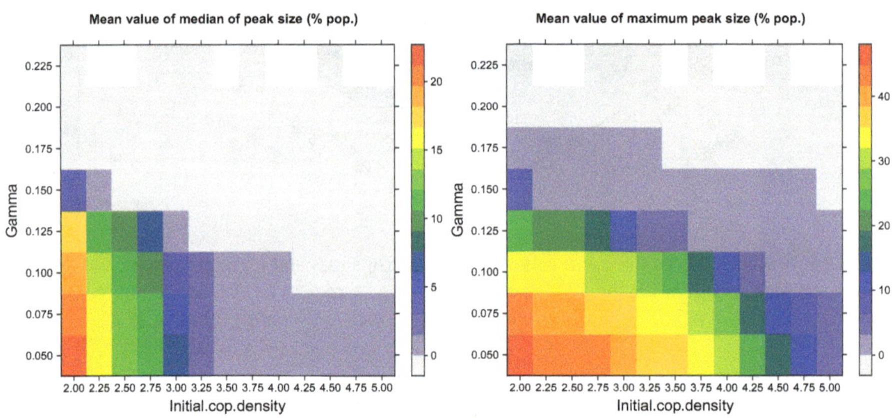

Fig. 6.28 Level plots of the mean values of the median and maximum peak size (% of the population) of rebellion bursts for the RD experiment, for the "large society" case, computed from 20 simulations for each pair of values of the sweeping parameters (Source: author)

sensitivity to deprivation with low legitimacy of the central authority, for the "small" and "large society" settings, under the *combined* effects of RD and legitimacy feedback.

The input parameters were the same described in the previous experiment (Table 6.2), except that `LF?` was set to *true*. The simulations were run considering homogeneous legitimacy perception (`LF-agents?` = *false*) and the interval for time moving average of the perceived legitimacy L_p was set to its default value (`L-memory` = 5 cycles).

Figures 6.34 and 6.35 show level plots of the % of the time with rebellion, and Figs. 6.36 and 6.37 level plots of the mean values of the median and maximum

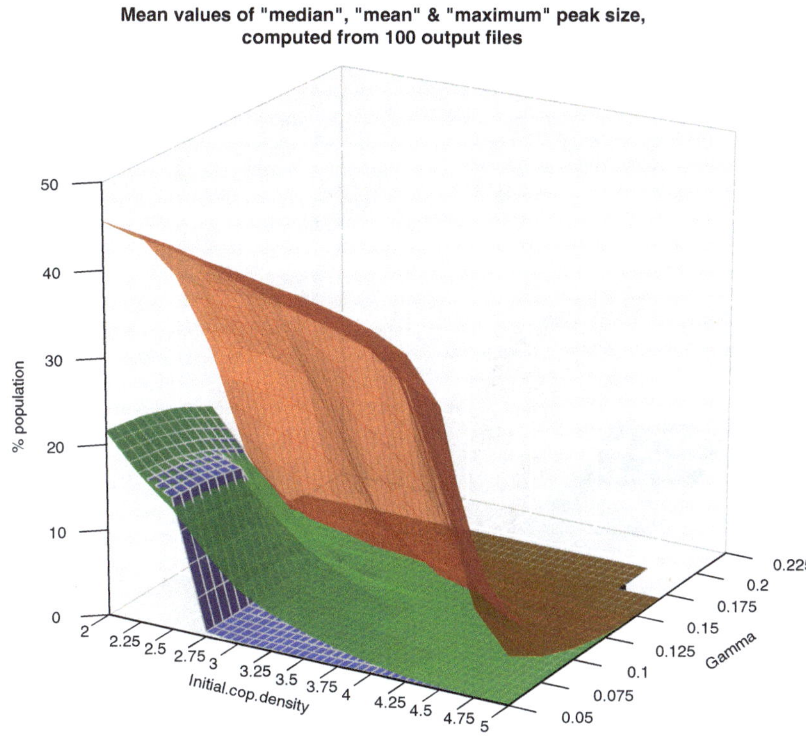

Fig. 6.29 Surface plots of mean values of the median (*blue*, opaque), mean (*green*, semi-transparent), and maximum (*red*, semi-transparent) peak size (% of the population) of rebellion bursts for the RD experiment, for the "small society" case, computed from 100 simulations for each pair of values of the sweeping parameters (Source: author)

peak size (% of the population) of the rebellion bursts, for the simulations in this experiment.

Comparing these with the corresponding figures in Sect. 6.3, it is apparent that, for the conditions analyzed in the experiment, sensitivity to deprivation is the dominant factor determining stability or instability. Introduction of legitimacy feedback does not change the qualitative behavior of the solutions. Although the % of time with rebellion and the peak sizes are larger when legitimacy feedback is combined with RD, the regions of stability, intermittent rebellion, and permanent instability in the level plots of % of time with rebellion and peak size are sensibly the same in both cases. In particular, the tipping point associated with γ is not shifted (at least significantly) by the introduction of legitimacy feedback.

It is interesting to examine in greater detail the time variations of legitimacy, for stable and unstable conditions. This can be done by comparing the time histories of the median of the perceived legitimacy for two simulations of the "large society" case with `initial-cop-density` = 4% and $\gamma = 0.10$ and $\gamma = 0.15$.

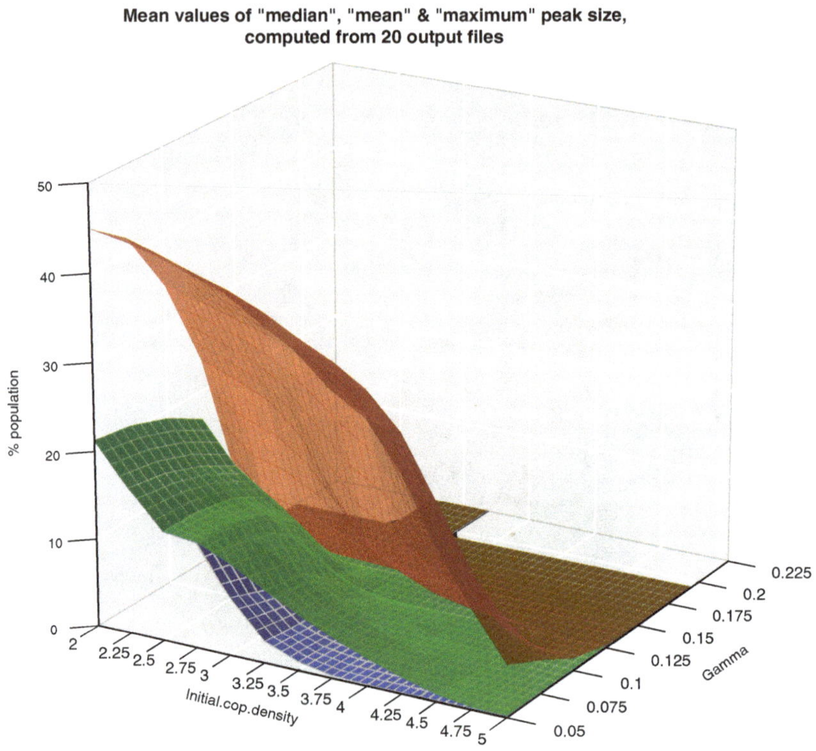

Fig. 6.30 Surface plots of mean values of the median (*blue*, opaque), mean (*green*, semi-transparent), and maximum (*red*, semi-transparent) peak size (% of the population) of rebellion bursts for the RD experiment, for the "large society" case, computed from 20 simulations for each pair of values of the sweeping parameters (Source: author)

According to Figs. 6.35 and 6.37, there is a change of qualitative behavior between these two conditions.

Figure 6.38 shows the time variation of the median of legitimacy (over all "citizens" and the "memory" length `L-memory = 5`) during the first two thousand cycles for five different runs of the two cases mentioned above. For $\gamma = 0.15$ the society is in a situation of stable calm and the legitimacy remains almost constant, with only small drops. However, for $\gamma = 0.10$ (larger sensitivity to value) the society is unstable. The legitimacy oscillates due to bursts of rebellion, which result in a large proportion of "citizens" turning "active" and being jailed by the "cops." This leads to the legitimacy drops observed.

When compared with the variations of the legitimacy indicators in, e.g., Fig. 4.12, it can be concluded that the legitimacy drops are larger for the ABM solution (maximum variation of 39%) than for the FSI "Legitimacy of the State" indicator. The oscillations of the simulated legitimacy perception are not present in the FSI indicators for the eight countries considered, but these have very slow variation (one value per year) and so cannot be put in realistic correspondence with the model.

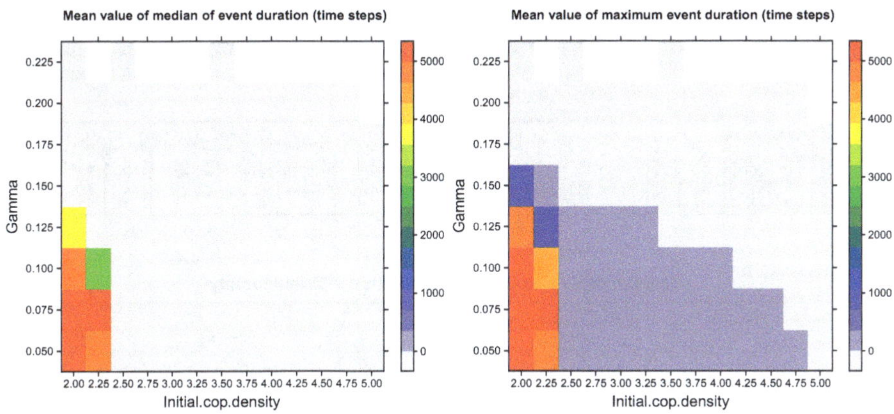

Fig. 6.31 Level plots of the mean values of the median and maximum event duration (in time steps) for the RD experiment, for the "small society" case, computed from 100 simulations for each pair of values of the sweeping parameters (Source: author)

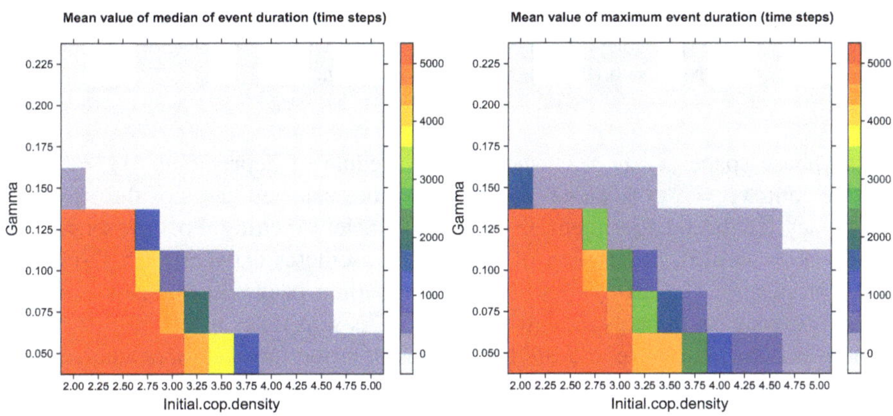

Fig. 6.32 Level plots of the mean values of the median and maximum event duration (in time steps) for the RD experiment, for the "large society" case, computed from 20 simulations for each pair of values of the sweeping parameters (Source: author)

Unstable solutions have the property of sensitivity to initial conditions. This is shown in Fig. 6.39, which illustrates the time variation of the median of legitimacy for the first five hundred cycles in five different simulations of the "large society" setting for `initial-cop-density` = 4% and $\gamma = 0.10$.

* * *

One interesting way of showing how RD-dependent hardship and legitimacy feedback lead to complex behavior is to analyze the trajectories of selected dependent variables in a suitable phase space. Figure 6.40 shows the trajectories

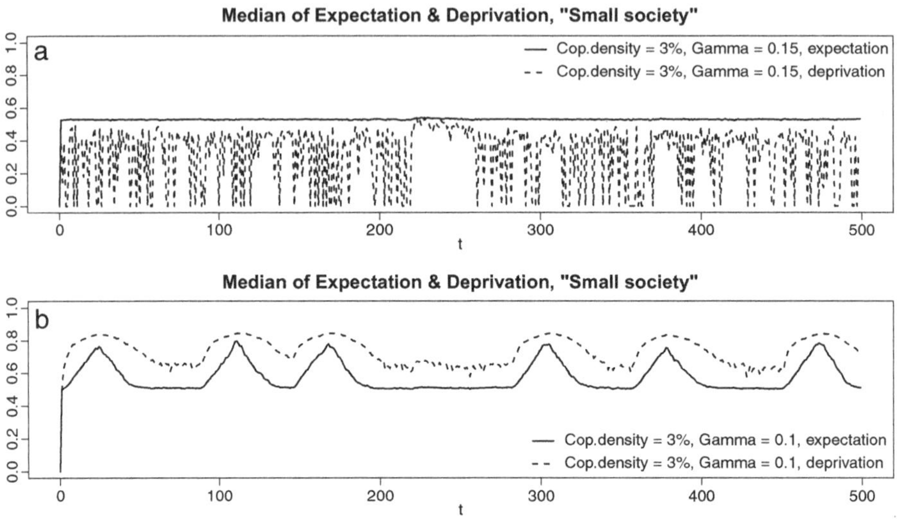

Fig. 6.33 Time history of the first 500 cycles of the medians of expectation and deprivation (over all "citizens") for the simulations of the "small society" case with `initial-cop-density` = 3.5%, for the two values $\gamma = 0.10$ and $\gamma = 0.15$ (Source: author)

in the phase space (x_1, x_2, x_3), where x_1 = median of deprivation,[5] x_2 = % of "active" and x_3 = % of "jailed" citizens, for four selected cases of the RD with legitimacy feedback experiment, in the "small society" setting. For low sensitivity to RD ($\gamma = 0.125$), the median of deprivation oscillates between 0.05 and 0.5, a small proportion of "citizens" is jailed, but rebellion peaks do not occur ("cops" arrest "citizens" as soon as they turn "active"). As sensitivity to RD increases, the society becomes unstable and the trajectories describe more complicated shapes. Increasing sensitivity to RD leads to higher levels of deprivation, but the maximum proportions of "active" and "jailed" citizens do not change significantly.

Figure 6.41 shows the trajectories in the phase space (x_1, x_2, x_3), where x_1 = median of deprivation, x_2 = % of "active" and x_3 = median of legitimacy,[6] for the same simulations ("small society" setting). Although the form of the trajectories is different, the results can be interpreted in a similar way: once the sensitivity to RD crosses a tipping point the artificial society becomes unstable; the median of deprivation increases with the sensitivity to RD, but the magnitudes of the legitimacy drops and the % of "active" citizens remain essentially unaltered for different cases of unstable solutions.

Figures 6.42 and 6.43 show the trajectories in the phase spaces considered above, for the same pairs of values of the sweeping parameters, but for the "large society"

[5] Here, the median is that of the distribution of the citizens' individual deprivation.

[6] In this case the value of the perceived legitimacy is identical for all agents and the median coincides with this value, because homogeneous legitimacy feedback was used in the simulations.

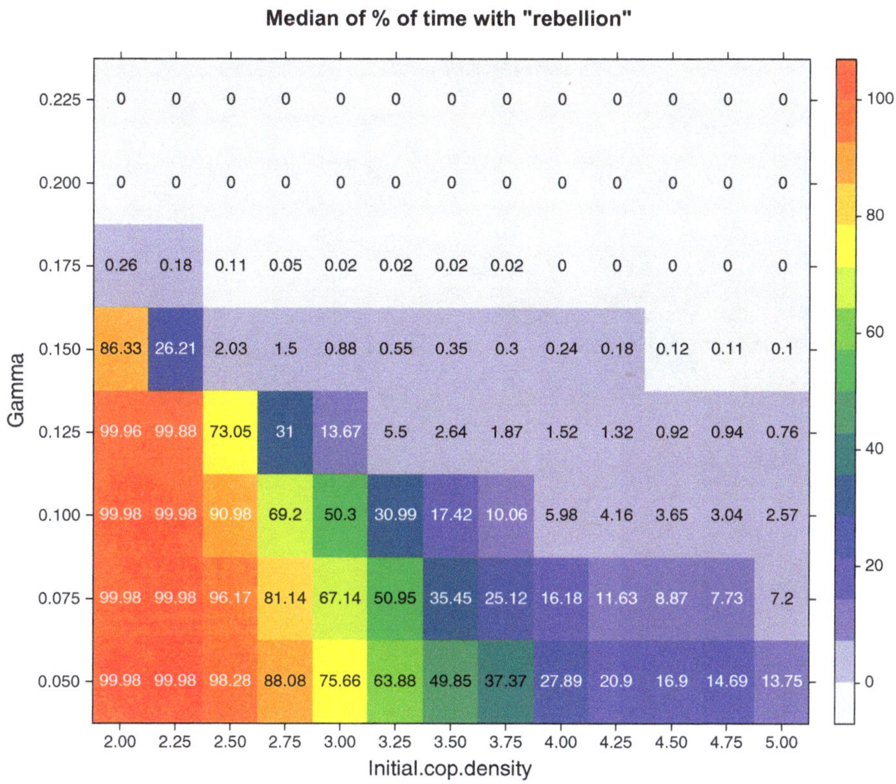

Fig. 6.34 Level plot of the median of the % of time with rebellion for the RD with legitimacy feedback experiment, for the "small society" case, computed from 100 simulations for each pair of values of the sweeping parameters (Source: author)

setting. The shape of the phase portraits is similar to that of the "small society" case, but for the "large society" setting, the points appear to fill the surfaces bounded by the "curves" in Figs. 6.40 and 6.41, near which the points in the trajectories shown in these latter figures tend to concentrate.

6.4.2 *Legitimacy Feedback in a Scenario of High Legitimacy and Low Repression*

The purpose of this second experiment with legitimacy feedback was to show how the introduction of this mechanism introduces instability in an otherwise stable scenario of high initial legitimacy and low repression.

In this experiment, hardship was set $H \sim \mathscr{U}(0, 1)$ as in Epstein's original model. The grid size, initial "citizen" and "cop" densities, vision radii, ρ_c and $J_{\max}$ were

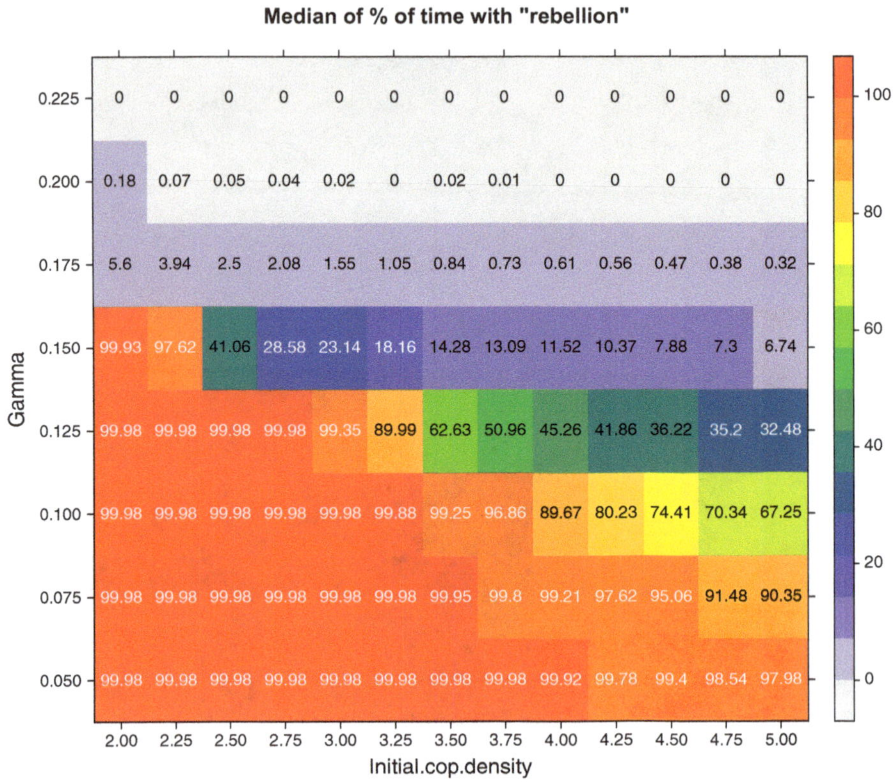

Fig. 6.35 Level plot of the median of the % of time with rebellion for the RD with legitimacy feedback experiment, for the "large society" case, computed from 20 simulations for each pair of values of the sweeping parameters (Source: author)

set to the default values shown in Table 5.4. Heterogeneous legitimacy feedback was introduced by setting $\texttt{LF?} = \textit{true}$ and $\texttt{LF} - \texttt{agents?} = \textit{true}$. Dependence of grievance on RD and network influences were not considered.

The initial legitimacy and threshold were set $L = 0.872$ and $T = 0.1$, respectively. For this value of the threshold, the legitimacy used in the simulations is only 0.028 below the value for which no "citizens" would turn active (0.9). Twenty simulations were run, each with a duration of 10,000 cycles.

Figure 6.44 shows the time history of the median of the citizens' perceived legitimacy L_p obtained in four simulations, which were selected to show the diversity of outcomes from this experiment.

In simulation 20, the solution remained stable for 10,000 cycles, with fluctuations of the median of L_p smaller than 2% relative to the initial value.

In simulation 7, the solution shows three small and short peaks of activity, which lead to (negative) legitimacy fluctuations of 11.5%, 7.4%, and 3.0%.

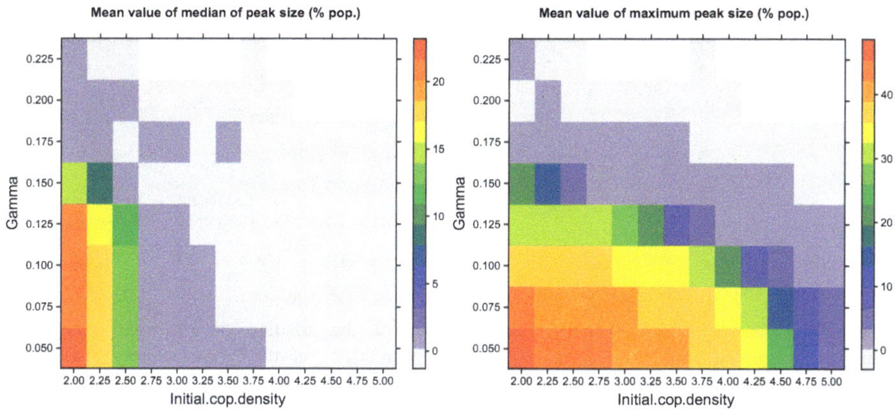

Fig. 6.36 Level plots of the mean values of the median and maximum peak size (% of the population) of rebellion bursts for the RD experiment with legitimacy feedback, for the "small society" case, computed from 100 simulations for each pair of values of the sweeping parameters (Source: author)

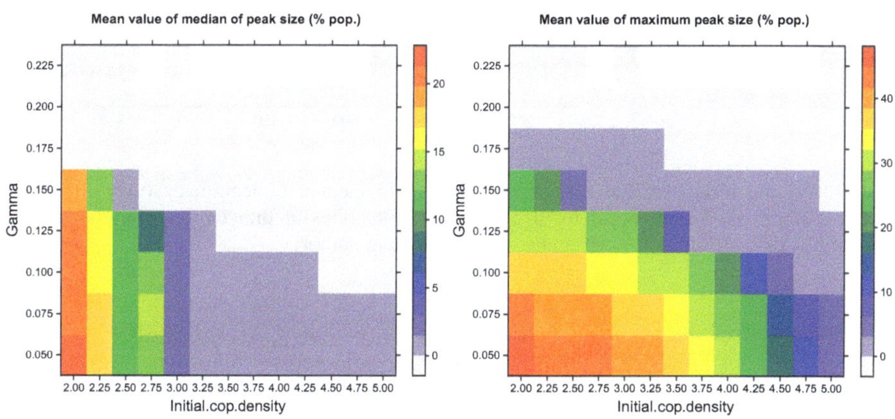

Fig. 6.37 Level plots of the mean values of the median and maximum peak size (% of the population) of rebellion bursts for the RD experiment with legitimacy feedback, for the "large society" case, computed from 20 simulations for each pair of values of the sweeping parameters (Source: author)

In simulation 11 there was one very large rebellion burst (with peak at cycle $t = 1864$) which lead to a legitimacy drop of 61% relative to the initial value, but for the rest of the simulation the artificial society remained stable with fluctuations of L_p smaller than 6%.

In simulation 3, the solution shows intermittent regime, with long periods of calm interrupted by five periods of turmoil lasting for hundreds of cycles, with large oscillating drops of L_p (maximum of 63.5% relative to the initial value), and one large isolated event (with peak at cycle $t = 5977$, with legitimacy drop of 63.2%).[7]

[7]It is not possible to determine whether the event with peak at $t = 9966$ at the end of the simulation corresponds to an isolated event or to a longer period of unrest.

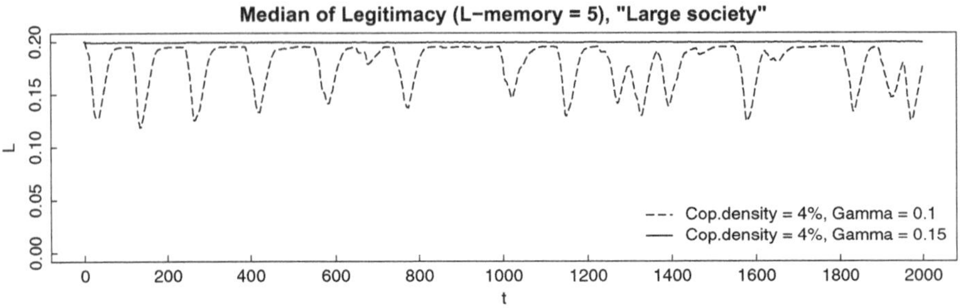

Fig. 6.38 Time history of the first 2000 cycles of the median of legitimacy (over all agents and `L-memory` = 5 time steps) for the simulations of the "large society" case with `initial-cop-density` = 4%, for the two values $\gamma = 0.10$ and $\gamma = 0.15$ (Source: author)

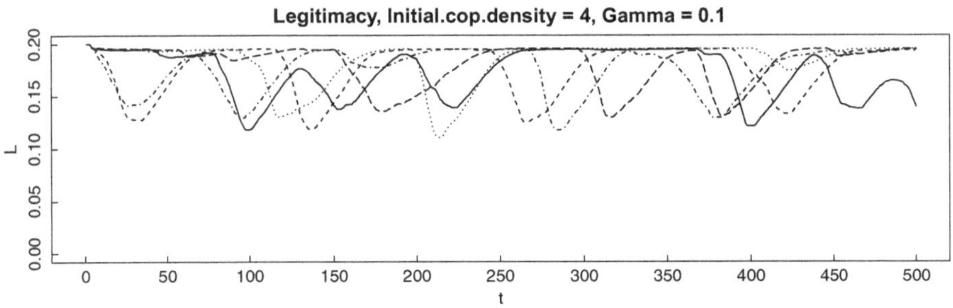

Fig. 6.39 Time history of the first 500 cycles of the median of legitimacy (over all agents and `L-memory` = 5 time steps) for five different simulations of the "large society" case with `initial-cop-density` = 4% and $\gamma = 0.10$ (Source: author)

Figure 6.45 shows the trajectories in the phase space (x_1, x_2), with $x_1 = \%$ of "active" citizens and x_2 = median of L_p, for the selected simulations.

For simulation 20 the trajectory reduced to small deviations from the equilibrium point $(x_1, x_2) = (0, 0.872)$. For simulation 7 the deviations are larger, but the trajectory remained confined to a small region near the equilibrium point.

The portrait for simulation 11 shows the effect of the large isolated event in phase space, leading to spanning of the trajectory over a wider region in the phase space. Points corresponding to the large event draw a sequence of points, starting at the equilibrium point with increasing x_1 and decreasing x_2, followed by a nearly horizontal trajectory with decreasing x_1 and nearly constant x_2, and a vertical trajectory with $x_1 = 0$ and increasing x_2 ending at the equilibrium point.[8] The trajectory for simulation 11 also includes a pattern of points concentrated near $(x_1, x_2) = (0, 0.872)$, which corresponds to small events like those in simulation 7.

[8]These correspond to the initial cascade of rebellion, "cops" jailing "active" citizens, and progressive release of the latter, respectively.

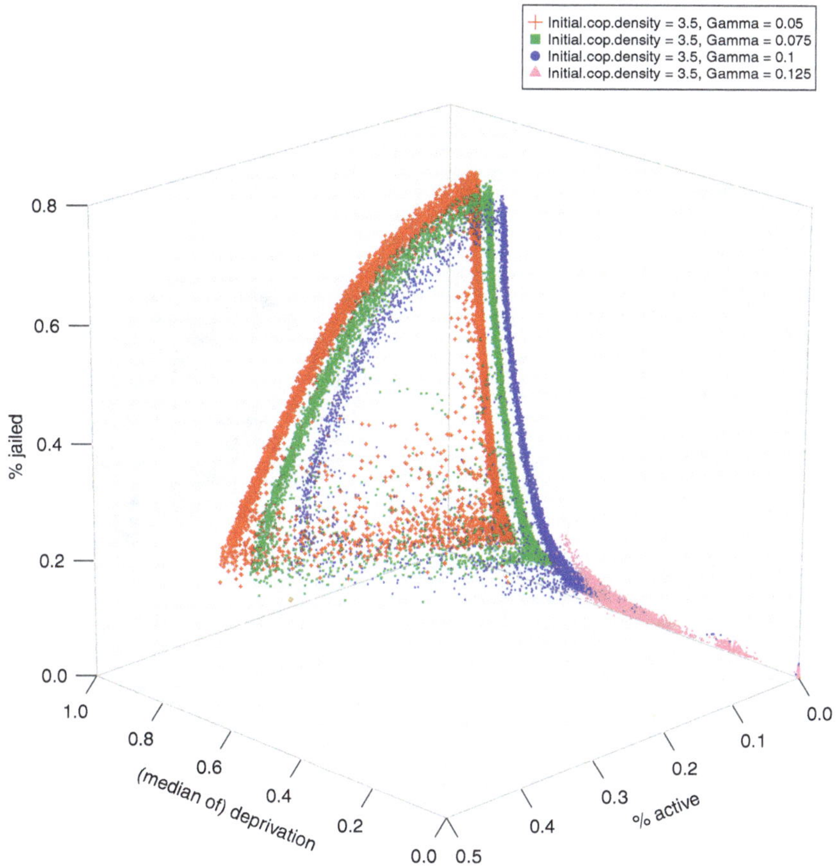

Fig. 6.40 Trajectories in the phase space of median of deprivation, % of "active" and % of "jailed" for four selected cases (pairs of values of the sweeping parameters) of the RD experiment with legitimacy feedback, "small society" case (Source: author)

The phase portrait of simulation 3 shows that the trajectory remained confined to a region of the phase space similar to that enclosed by the trajectory resulting from the (single) large rebellion event in simulation 11. However, the occurrence of many peaks of different magnitude led to points of the trajectories filling that region.

This is also the reason why the phase portraits for the "large society" setting in the previous experiment show trajectories filling regions of the phase space(s) instead of remaining concentrated near closed "curves" as in the "small society" case: the larger number of "citizens" increases the variance of peak sizes, which in turn leads to a greater multiplicity of possible trajectories.

In summary, the fact that the legitimacy feedback mechanism implemented in the model leads to solutions with such a varied and unpredictable behavior, for a value of initial legitimacy so close to the limit for which no "citizens" rebel, is remarkable and unexpected.

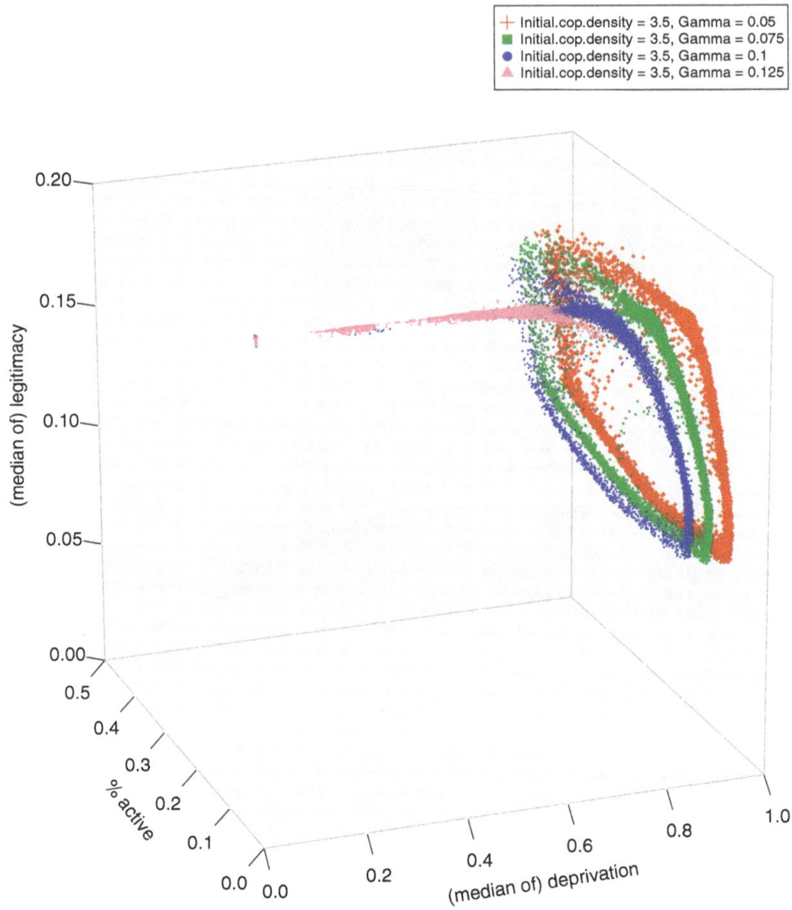

Fig. 6.41 Trajectories in the phase space of median of deprivation, % of "active" and median of legitimacy for four selected cases (pairs of values of the sweeping parameters) of the RD experiment with legitimacy feedback, for the "small society" case (Source: author)

6.5 Network Influences

Network influences are essential for describing social conflict phenomena. The widespread use of social media such as blogs and Facebook played a central role in the propagation of protests in recent conflict processes, particularly in the AS. Faris [25] and Comninos [13] discussed the use of SN in the AS and concluded that these are not the cause of social conflict and revolution but can trigger informational cascades that change the dynamics of the events.

In the "abstract" ABM proposed herein, networks can be interpreted as simplified representations of fixed but non-local information and influence structures (involving fixed sets of link neighbors of a "citizen"), whereas the cells within a citizen's vision radius can be thought of as a local-context "crowd" or "flock" influence space.

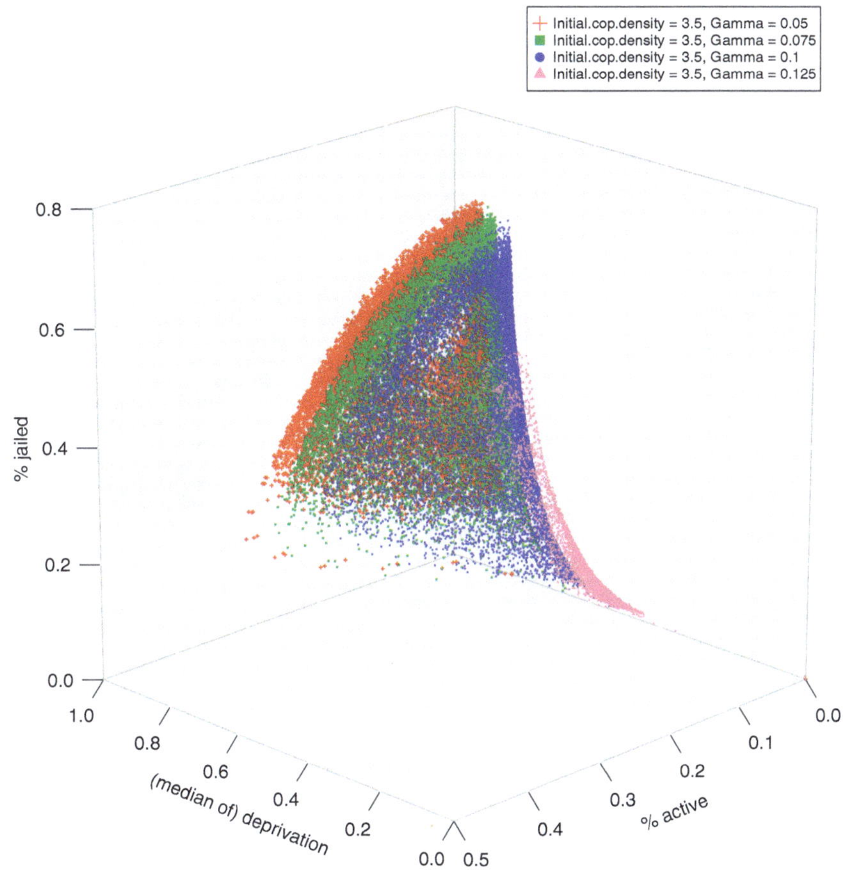

Fig. 6.42 Trajectories in the phase space of median of deprivation, % of "active" and % of "jailed" for four selected cases (pairs of values of the sweeping parameters) of the RD experiment with legitimacy feedback, "large society" case (Source: author)

The main purpose of the computer experiments described in this section was to evaluate how network influences change the solutions' behavior, for the context of low government legitimacy, RD-dependent hardship and legitimacy feedback considered before (Sects. 6.3–6.4). Another goal was to investigate the effects of varying the degree of connectivity and influence weights, and whether or not the two types of network ("group" and "infl") lead to substantially different magnitudes and durations of the simulated events.[9]

[9]In the "group" network, "citizens" have a fixed number of link neighbors, and since the network is undirected cyclic reinforcement is possible. In the "infl" network, the probability of a "citizen" being connected to one or more "activists" is an important factor.

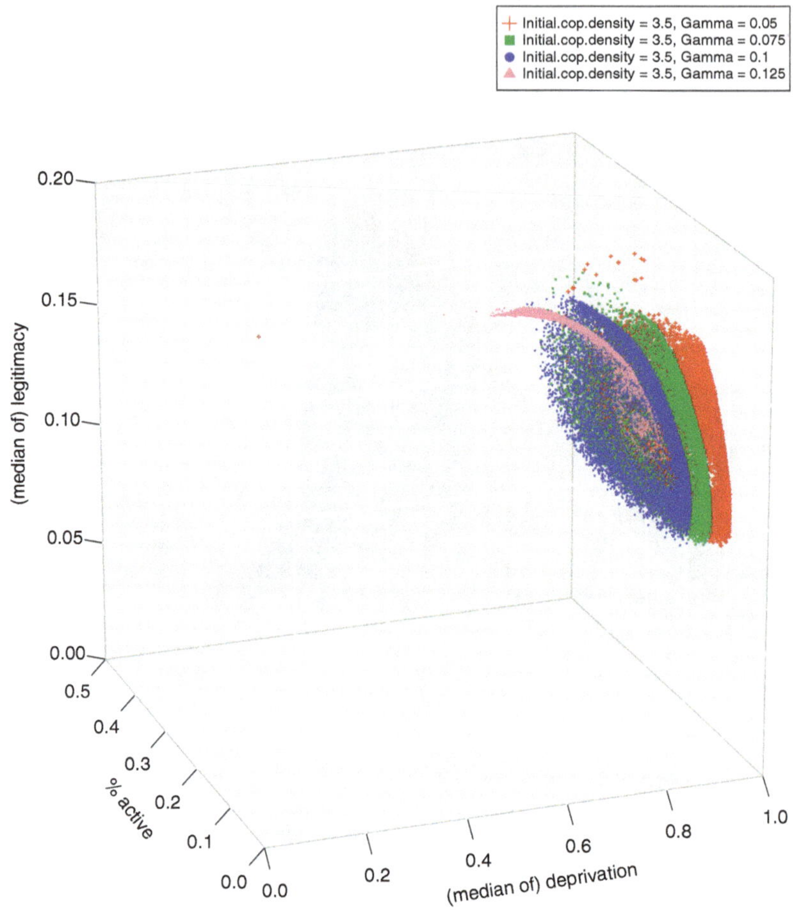

Fig. 6.43 Trajectories in the phase space of median of deprivation, % of "active" and median of legitimacy for four selected cases (pairs of values of the sweeping parameters) of the RD experiment with legitimacy feedback, for the "large society" case (Source: author)

To determine whether or not network influences induce instability, the case with parameter values $\gamma = 0.15$ and `initial-cop-density` $= 3.5\%$ described in Sect. 6.4 was chosen are reference. As shown in Figs. 6.34 and 6.36, this corresponds to stable conditions.

Two sets of experiments were performed, one with "group" and another with "infl" networks. Table 6.3 shows the values of the input parameters for the two sets of experiments. The model space was the default 40×40 2D torus in both "group" and "infl" experiments ("small society" setting). The number of "activist" citizens in the "infl" simulations was arbitrarily set to `num-infl` $= 10$.

The sweeping variables were the influence weights, `w-group` and `w-infl`, and the `group-size` and "audience" `infl-size` of each "activist," respectively.

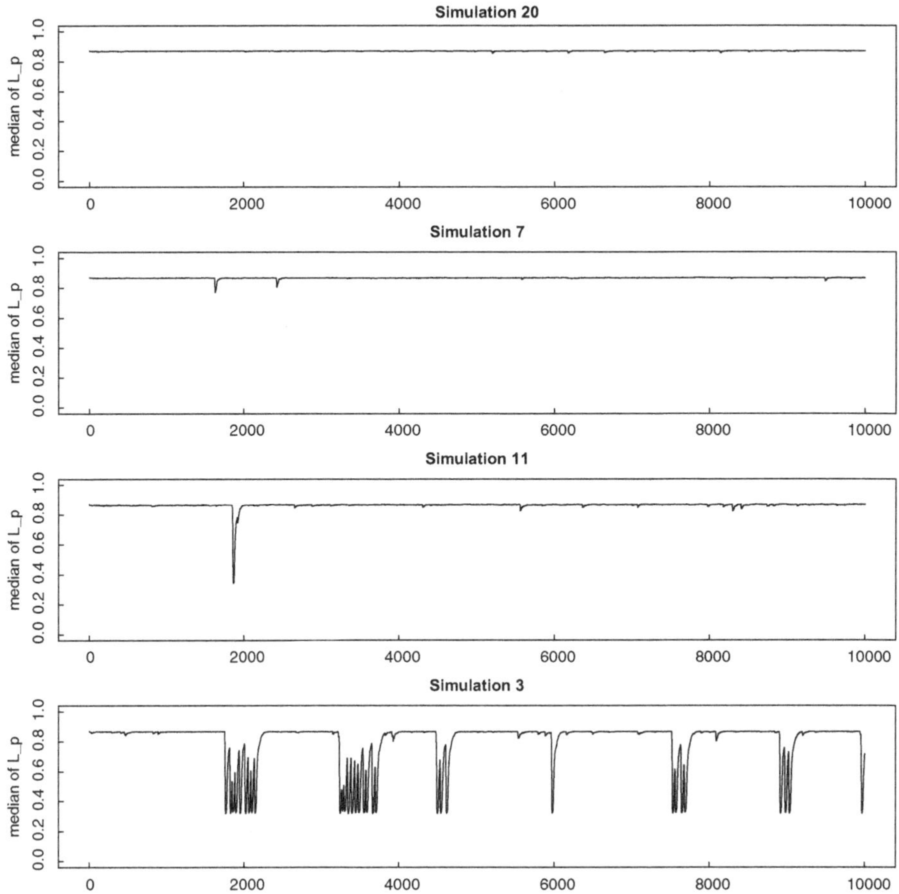

Fig. 6.44 Time history of the median of the (heterogeneous) perceived legitimacy in four selected simulations of the legitimacy feedback in high legitimacy and low repression scenario (Source: author)

The values of the influence weights were the same in both experiments. For the conditions of the "infl" experiment, the probability of one "citizen" being influenced by at least one "activist" varied between 0.40 and 0.97. One hundred simulations were performed for each pair of sweeping parameter values, and each simulation was run for 5,000 cycles.

Figures 6.46 and 6.47 show level plots of the mean values of the median % of the total time with rebellion, computed from 100 output files, for the simulations with the "group" and "infl" networks, respectively.

In the case of "group" network influences (Fig. 6.46), the median of the % of time with rebellion increased between $\sim 2.5\times$ and $\sim 11\times$ relative to the reference conditions (no network influences). For the values of the sweeping parameters

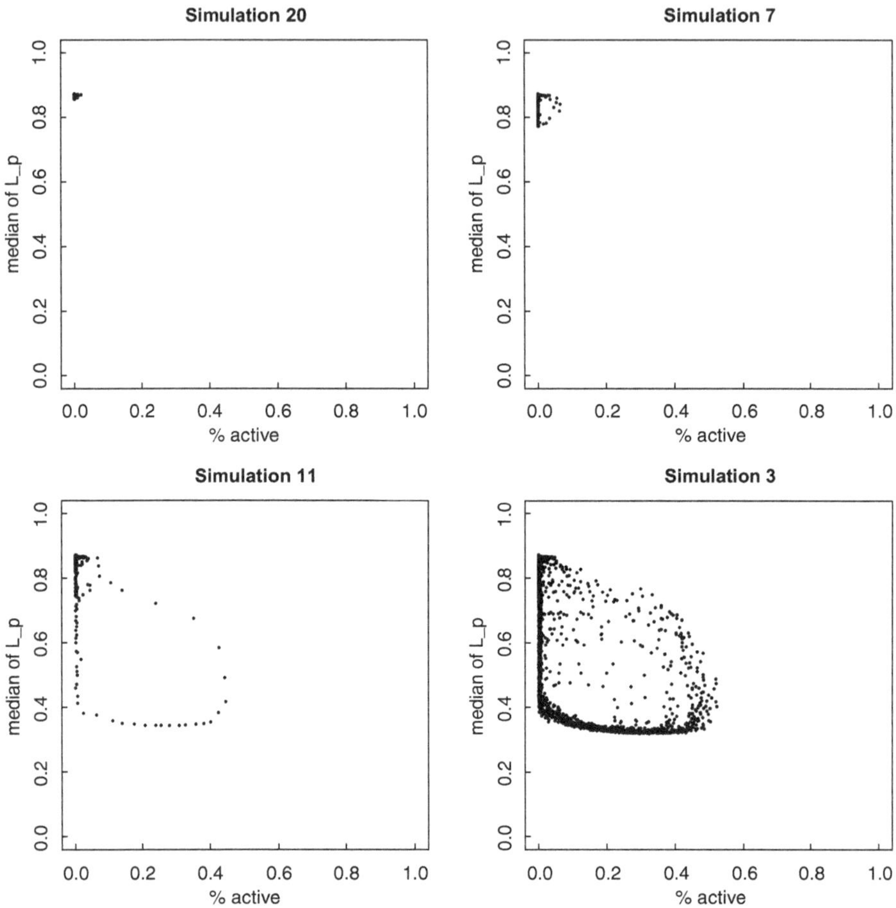

Fig. 6.45 Time history of the trajectories in the (x_1, x_2) phase space, with $x_1 == \%$ of "active" citizens and $x_2 =$ median of the (heterogeneous) perceived legitimacy, in four selected simulations of the legitimacy feedback in high legitimacy and low repression scenario (Source: author)

considered in this experiment, increments of both `group-size` and `w-group` lead to increased % of time with rebellion, although the variations of the latter are not monotonic with respect to both sweeping parameters.

In the case of the "infl" network (Fig. 6.47), the median of % of time with rebellion increased between $\sim 1.1\times$ and $\sim 2.8\times$ relative to the reference conditions (no network influences). Although the increase of % of time with rebellion was lower than in the previous case ("group" network influences), the results are qualitatively similar (increments of both `infl-size` and `w-infl` lead to increased % of time with rebellion).

Table 6.3 Input parameters for the simulations with the "group" and "infl" networks (Source: author)

	"group" experiment	"infl" experiment
Parameter name	Value(s)	Value(s)
`world-width`	40	40
`world-height`	40	40
`initial-cop-density`	3.5%	3.5%
`initial-citizen-density`	70%	70%
Number of "citizen" agents	1120	1120
Number of "cop" agents	56	56
`threshold`	0.60	0.60
`government-legitimacy`	0.20	0.20
`max-jail-term`	40	40
`RD?`	*true*	*true*
`gamma`	0.15	0.15
`LF?`	*true*	*true*
`GROUP?`	*true*	*false*
`group-size`	{5,6,7,8,9,10}	–
`w-group`	0.8 to 1.3, step 0.1	–
`INFL?`	*false*	*true*
`infl-size`	–	5% to 30%, step 5%
`num-infl`	–	10
`w-infl`	–	0.8 to 1.3, step 0.1

Figures 6.48 and 6.49 show level plots of the mean values of the median and maximum peak size (% of the population), computed from 100 output files, for the simulations with the "group" and "infl" networks, respectively.

In the case of "group" network influences the mean value of the median of peak size was residual and identical to that obtained for the reference simulation (no network influences). In the "infl" experiment the mean value of the median of peak size increased with both `infl-size` and `w-infl`, but also remained residual.

However, the mean values of the maximum peak size increased with the both the degree of connectivity (`group-size` and `infl-size`) and the influence weights (`w-group` and `w-infl`) in both experiments. For the largest values of `group-size` and `infl-size` considered, the maximum peak size attained very large values ($\sim$ 52% and $\sim$ 43% for the "group" and "infl" network influences, respectively).

Analysis of Figs. 6.48 and 6.49 also shows that for the ranges of the sweeping parameters considered, connectivity had a larger impact on the magnitude of the rebellion peaks than the influence weight.

Figures 6.50 and 6.51 show level plots of the mean values of the median and maximum event duration (in time steps), computed from 100 output files, for the simulations with the "group" and "infl" networks, respectively. These figures show that the variations of event duration with respect to both sweeping parameters follow

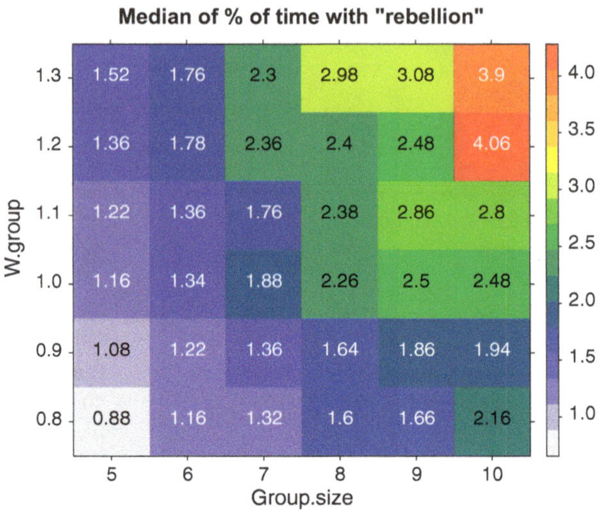

Fig. 6.46 Mean values of the median of the % of time with rebellion for the simulations of the network influence effects with the "group" network, computed from 100 simulations for each pair of values of the sweeping parameters (Source: author)

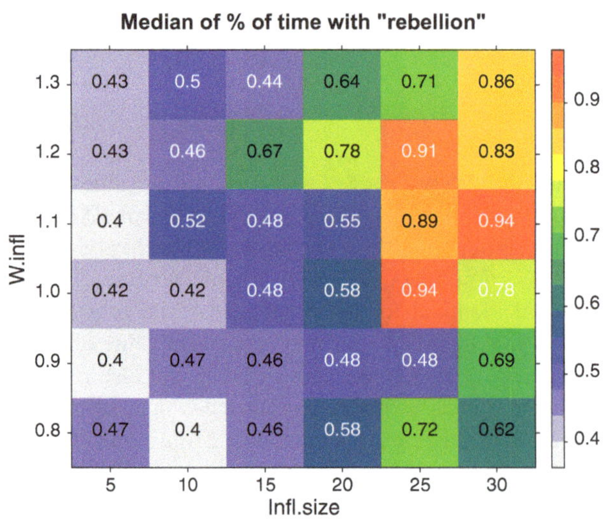

Fig. 6.47 Mean values of the median of the % of time with rebellion for the simulations of the network influence effects with the "infl" network, computed from 100 simulations for each pair of values of the sweeping parameters (Source: author)

the same patterns as the variations of peak size previously illustrated in Figs. 6.48 and 6.49. This is due to the fact that larger outbursts also take longer for "cops" to suppress them.

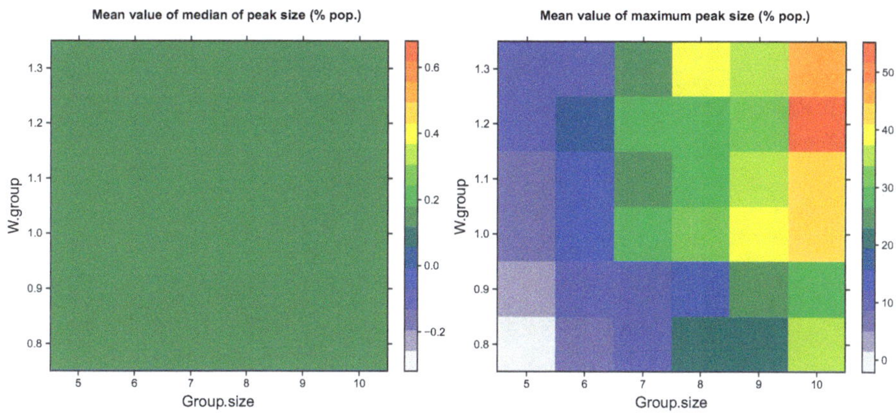

Fig. 6.48 Level plots of the mean values of the median and maximum peak size (% of the population) of rebellion bursts for the simulations of the network influence effects with the "group" network, computed from 100 simulations for each pair of values of the sweeping parameters (Source: author)

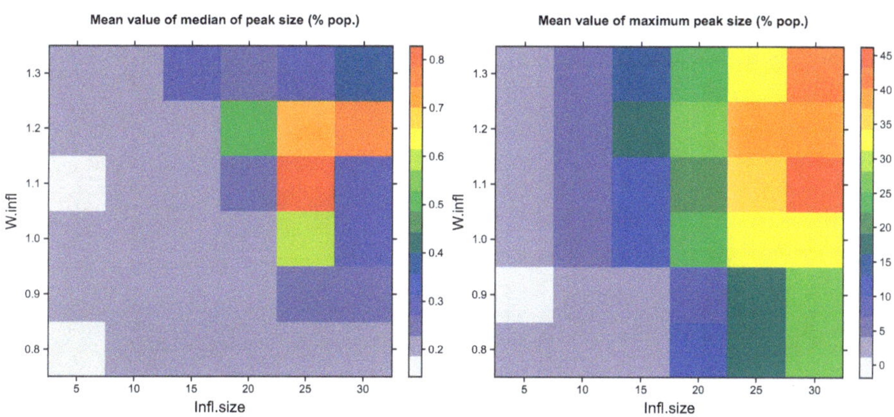

Fig. 6.49 Level plots of the mean values of the median and maximum peak size (% of the population) of rebellion bursts for the simulations of the network influence effects with the "infl" network, computed from 100 simulations for each pair of values of the sweeping parameters (Source: author)

In summary, the present model of network influences leads to instability of otherwise stable solutions for both "group" and "infl" ("activists") networks. The network influences have small effect on the % of time with rebellion as well as the peak size and duration of most events, but greatly amplify isolated events. This is illustrated in Fig. 6.52, which shows the simulated time history of the % of "active" citizens for four runs of the "group" network influence experiment with `group-size` = 9 and `w-group` = 0.9.

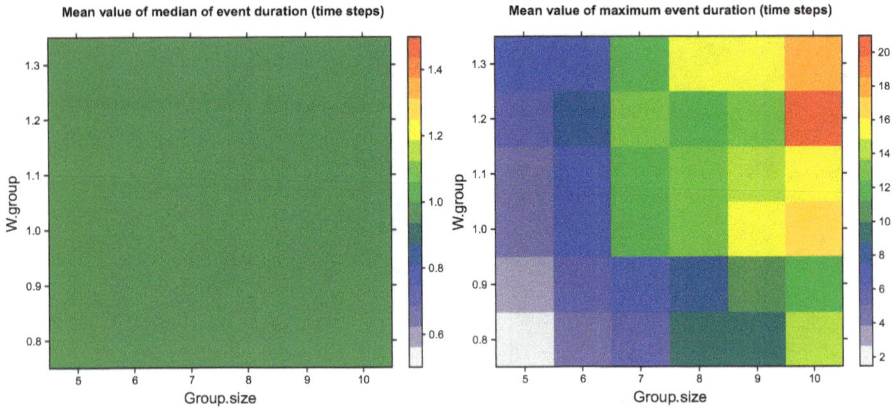

Fig. 6.50 Level plots of the mean values of the median and maximum event duration (in time steps) of rebellion bursts for the simulations of the network influence effects with the "group" network, computed from 100 simulations for each pair of values of the sweeping parameters (Source: author)

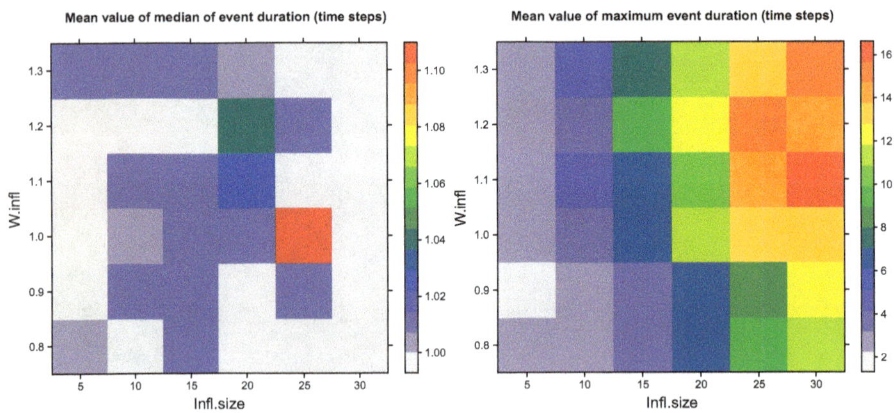

Fig. 6.51 Level plots of the mean values of the median and maximum event duration (in time steps) of rebellion bursts for the simulations of the network influence effects with the "infl" network, computed from 100 simulations for each pair of values of the sweeping parameters (Source: author)

For both types of networks, the destabilizing effect increases with both the influence weight and degree of connectivity, but the latter has a larger impact on the magnitude of the simulated conflict events than the influence weight.

In future experiments, it would be interesting to repeat the simulations described in this section with smaller increments and wider ranges of the sweeping parameters. It would also be important to run simulations for the "large society" setting considered in Sects. 6.3–6.4. In the case of the "infl" network it would also be interesting to vary the number of "activists." Since network influences are a key feature in conflict dynamics, it would also be important to study their destabilizing effect independently of legitimacy variations and sensitivity to deprivation.

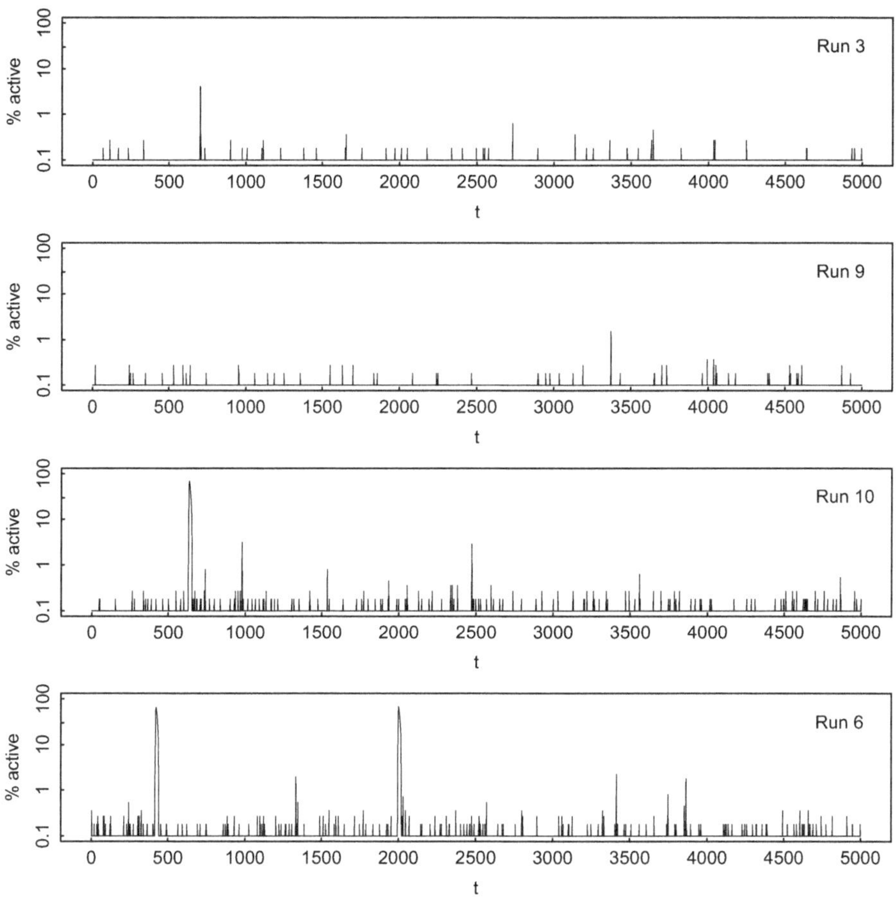

Fig. 6.52 Simulated time history of the % of "active" citizens obtained in four different simulations of the "group" network influence experiment, for `group-size` = 9 and `w-group` = 0.9 (Source: author)

Modeling network influences in ABM of social conflict is a very important and complicated subject, which requires a deeper study than the one performed in the present work. This involves the analysis and development of different formulations for the social influence (e.g., nonlinear forms of dispositional contagion), consideration of different network sizes and topologies, and information based on real data for guiding parameterization, exploration and discussion of the improved model(s).

Chapter 7
Conclusions and Future Work

In this work, the dynamics of politically motivated large-scale conflict against a central authority was studied using an ABM of "abstract" type, and the results of the computer simulations were discussed using analyses of conflict events in eight African countries affected by the AS. The leitmotif was the idea of extending Epstein's classical ABM of civil violence by combining four elements: (*i*) theories of social conflict; (*ii*) analysis of datasets of social indicators and conflict events; (*iii*) the role and meaning of the model parameters; and (*iv*) emphasis on how the mechanisms described in theories can be represented in an ABM.

7.1 Conclusions

The analysis of eight African countries affected by the AS (Algeria, Egypt, Libya, Mali, Mauritania, Morocco, Sudan, and Tunisia) combining the SCAD database of conflict events, and FSI, FWI and Gini indicators, provided important insights on the AS itself and on the significance of complexity for studying large-scale conflict processes. It also yielded plausible estimates for ABM parameterization and for the magnitude, duration, and recurrence of simulated conflict events. The salient results of this analysis can be summarized as follows:

- There were massive demonstrations and riots before the beginning of the AS in several of the African countries analyzed. The AS greatly increased the number and frequency of conflict events, particularly demonstrations;
- Spontaneous demonstrations (and riots) were more numerous than organized events. Escalation to riots in initially peaceful demonstrations was significant, but riots did not evolve into more violent conflict manifestations (e.g., anti-governmental violence);
- The % of the time with calm and activity is important for comparing with ABM simulations, because it is independent of the time scale. For demonstrations,

C.M. Lemos, *Agent-Based Modeling of Social Conflict*, SpringerBriefs in Complexity, https://doi.org/10.1007/978-3-319-67050-8_7

the characteristic % of time with activity was 14% (Egypt)–16% (all countries) before and 26% (Egypt)–57% (all countries) after the beginning of the AS. For riots, the corresponding values were 4% (Egypt)–4% (all countries) before and 3% (Egypt)–7% (all countries) after the beginning of the AS;

- In general demonstrations were short events, with median one day and in 75% of the cases they were shorter than ten days. The time interval between events was very heterogeneous among the eight countries, particularly before the beginning of the AS, for both demonstrations and riots.

The innovations of the proposed ABM are:

- **Analysis of the "citizens" decision rule**: The analytical study of the pdf of $G-N$ in the "citizens" decision rule in Epstein's ABM provided an explanation of why some previously proposed forms the estimated arrest probability function lead to large rebellion peaks whereas others do not. It was shown that large rebellion peaks are possible if the estimated arrest probability drops to zero below a critical value of the ratio ρ_c of the number of visible "cops" to visible "active" citizens, and a form of the estimated arrest probability with ρ_c as a controlling parameter was proposed. A relationship was derived between the threshold T, the legitimacy L, and the estimated arrest probability $P_a(\rho)$ for significant rebellion peaks to occur;
- **Extension of the "citizens" decision rule**: The decision Rule A was extended to include network influences via the mechanism of *dispositional contagion*. This formulation was inspired in Ted Gurr [38] and Epstein [23];
- **RD (economic & political)**: The model describes hardship as a function RD which incorporates two key propositions (I.2 and I.4) of Ted Gurr's frustration-aggression theory. This formulation is based on: (*i*) setting a Pareto-distributed `value` attribute of "citizen" agents; (*ii*) computing RD as the difference between the median of "value" of visible "citizens" and the citizen's own "value;" and (*iii*) using a power function to model the emotional intensity of the response to the distance (gap) between expectation and own "value." The exponent γ of the power function controls sensitivity to value and can be used to represent the difference between political and economic RD;
- **Legitimacy feedback**: The mechanism of (endogenous) legitimacy feedback was implemented using a simple linear function based on Gilley's theoretical framework on the measurement of legitimacy [33, 34], combined with memory effects (to represent a delay between bursts of rebellion and legitimacy drops). Two different forms of the legitimacy feedback mechanism were implemented, homogeneous (global) and heterogeneous (variable among "citizens");
- **Network Influences**: Two different networks were considered: (*i*) "group" represented by a union of small undirected cliques, and (*ii*) "infl" (for "influentials") represented by a union of directed star networks with randomly chosen "citizen" agents as hubs. These two networks provide an abstract representation of two important influence modes in a society, one associated with highly cohesive small scale communities connected by strong undirected links (two-way influence), and another associated with (weaker, one-way) directed links through which influential agents ("activists") shape global perceptions.

The model was explored in a series of computer experiments for studying the influence of parameter values, effects and mechanisms on the qualitative and quantitative properties of the solutions. The most salient findings were:

- The parameter ρ_c in the estimated arrest probability strongly influences the magnitude of the simulated conflict events and has associated tipping points. The maximum jail term $J_{\max}$ strongly influences the interval between successive events, and no tipping points associated with $J_{\max}$ were found;
- The model can produce solutions with three distinct regimes or long-term behaviors—permanent calm, complex (i.e., with intermittent peaks of rebellion) and permanent turmoil—if inequality (6.5) (page 74) is used to set suitable values for the threshold T. This allowed simulations with three different regimes (permanent calm, punctuated equilibrium, and permanent rebellion) for low values of legitimacy (characteristic of authoritarian regimes), instead of the high values of legitimacy used in previous works (e.g., [21, 24, 26, 52]);
- The system's stability is much more sensitive to the parameter γ (sensitivity to RD) than to the deterrence capability of the central authority, and it was found that γ has an associated tipping point. This provides a plausible model representation for theories stating that political deprivation is more important than economic deprivation as a source of large-scale social conflict [9, 66], and that once the sensitivity to RD surpasses a certain limit there will be social unrest, even for strong levels of repression (as suggested by Gene Sharp's theory of non-violent action [66]);
- When legitimacy feedback is combined with RD-dependent hardship, the solutions show the same three regimes as in the simulations of the RD-dependent hardship, for the same ranges of γ and "cop" density. Thus, if instability is triggered by strong RD, endogenous legitimacy feedback does not change the qualitative behavior of the solutions. When the solutions are plotted in the two 3-D phase spaces {RD, % "active", % "jailed"} and {RD, % "active", L_p}, it is observed that for certain combinations of the sweeping input parameters the orbits describe trajectories with complicated shapes. Legitimacy feedback induces instability, and under certain conditions of high legitimacy and low level of repression leads to solutions with alternating periods of calm and turmoil. Observation of the simulated time series of legitimacy shows sensitivity of the solutions to the initial conditions;
- In the present formulation based on dispositional contagion and linear superposition, network influences due to both the "group" and "infl" networks introduce instability, by greatly amplifying the magnitude and increasing the duration of simulated conflict events in otherwise stable conditions. Instability increases with both the influence weights and the degree of connectivity, but the degree of connectivity has a larger impact than the influence weight on the magnitude of the simulated rebellion events .

Figure 7.1 summarizes the main contributions of the present work in graphical form.

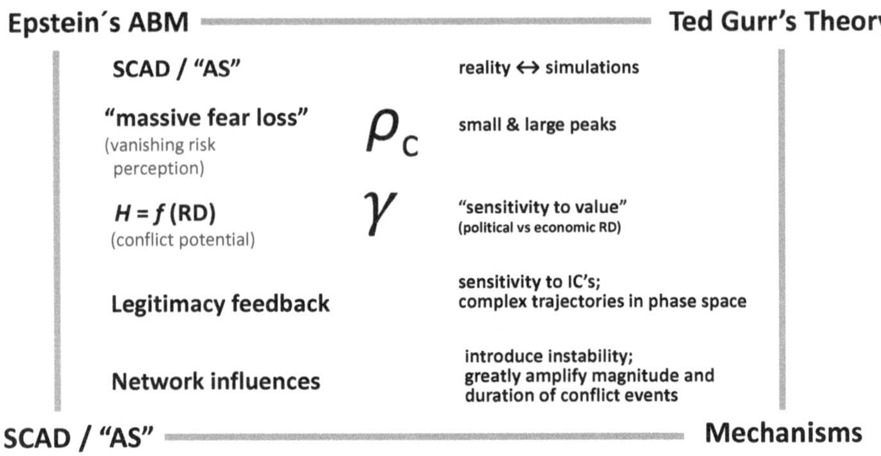

Fig. 7.1 Summary of contributions of the present work (Source: author)

7.2 Future Work

The present work can be improved in several different ways. One of these is the further exploration of the model for time-varying expectation, deprivation, and deterrence capability, following the approach described in [21, 24]. For instance, it would be interesting to determine the model's ability to simulate the J-curve and other deprivation patterns (Fig. 2.1).

Another possible improvement is the development of a more refined model of network influences, considering the size and structure of the networks, non-linear forms of dispositional contagion, and eventually memory effects. This would require a thorough theoretical review encompassing the size and structure of the networks to be used for realistic simulations. Also, it would be important to set the size and structure of the networks based on data from real situations. However, the relationship between the networks' sizes (e.g., a typical family size or the number of "activists" and their audiences) and population sizes is difficult to reproduce in models.

Still another important development of great practical interest would be the extension of the present model to the case of ethnic and religious conflict. Ethnic and religious conflicts involve social identity [68] and social RD, which arises from comparison between in-group and out-group status and capabilities [38]. The fundamental mechanisms of deterrence and group support are present, but deterrence depends on both "cops" and members of the rival ethnic. Since these conflicts tend to be more violent than those considered in this work, it would be necessary to model killing of agents by the rival ethnic and introduce some form of population dynamics [21, 24].

References

1. K. Allan, *The Social Lens: An Invitation to Social and Sociology Theory* (SAGE Publications, Thousand Oaks, 2007)
2. F. Amblard, A. Geller, M. Neumann, A. Srbljinovic, N. Wijermans, *NATO Science for Peace and Security Studies - E: Human and Societal Dynamics*, volume Complex Societal Dynamics - Security Challenges and Opportunities, chapter Analyzing Social Conflict via Computational Social Simulation: A Review of Approaches (IOS Press, Amsterdam, 2010), pp. 126–141. NATO
3. S.E. Asch, *Effects of Group Pressure on the Modification and Distortion of Judgements* (Carnegie Press, Pittsburg, PA, 1951), pp. 177–190
4. R. Axelrod, The dissemination of culture: a model with local convergence and global polarization. J. Confl. Resolut. **41**, 203–226 (1997)
5. P. Baker, *The Conflict Assessment System Tool (CAST)* (The Fund for Peace, New York, 2006)
6. D. Beetham, Max weber and the legitimacy of the modern state. Analyse Kritik **13**, 34–45 (1991)
7. L. Berkowitz, Frustration-aggression hypothesis: examination and reformulation. Psychol. Bull. **106**(1), 59–73 (1989)
8. T.A. Birkland, *An Introduction to the Policy Process: Theories, Concepts, and Models of Public Policy Making*, 3rd edn. (Routledge, Abingdon, 2015)
9. D. Bischof, Why Arabs Rebel - Relative Deprivation Revisited. Master's thesis, Fakultät Sozial und Wirtschaftswissenschaften der Otto-Friedrich-Universität Bamberg, 2012
10. R. Collins, *Conflict Sociology: Toward an Explanatory Science* (Academic Press, New York, 1975)
11. R. Collins, *Violence: A Micro-sociological Theory* (Princeton University Press, Princeton, 2008)
12. R. Collins, Micro and macro causes of violence. Int. J. Conflict Violence **3**(1), 9–22 (2009)
13. A. Comninos, Twitter revolutions and cyber crackdowns. User-generated content and social networking in the Arab spring and beyond. Technical report, Association of Progressive Communication (PAC), 2011
14. L.A. Coser, Social conflict and the theory of social change. Br. J. Sociol. **8**(3), 197–207 (1957)
15. R. Dahrendorf, Toward a theory of social conflict. J. Confl. Resolut. **2**(2), 170–183 (1958)
16. R. Dahrendorf, *Class and Class Conflict in Industrial Society* (Stanford University Press, Stanford, 1959)
17. J.C. Davies, Toward a theory of revolution. Am. Sociol. Rev. **27**(1), 5–19 (1962)
18. T.P. Davies, H.M. Fry, A.G. Wilson, S.R. Bishop, A mathematical model of the London riots and their policing. Sci. Rep. **3**(1303), (February 2013)

C.M. Lemos, *Agent-Based Modeling of Social Conflict*, SpringerBriefs in Complexity, https://doi.org/10.1007/978-3-319-67050-8

19. J. Dollard, L.W. Doob, N.E. Miller, O.H. Mowrer, R.R. Sears, *Frustration and Aggression* (Yale University Press, New Haven, 1939)
20. J. Doran, Iruba: An Agent-Based Model of the Guerrilla War Process. In: Troitzsch, K. (ed.), *Pre-Proceedings of the Third Conference of the European Social Simulation Association, Koblenz, September 5–9*, pp. 198–205, September 2005
21. J.M. Epstein, Modeling civil violence: An agent-based computational approach. Proc. Natl. Acad. Sci. USA **99**, 7243–7250 (2002)
22. J.M. Epstein, *Generative Social Science: Studies in Agent-Based Computational Modeling* (Princeton University Press, Princeton, 2006)
23. J.M. Epstein, *Agent_Zero. Toward Neurocognitive Foundations for Generative Social Science* (Princeton University Press, Princeton, 2013)
24. J.M. Epstein, J.D. Steinbruner, M.T. Parker, Modeling Civil Violence: An Agent-Based Computational Approach. Center on Social and Economic Dynamics, Working Paper No. 20, January 2001
25. D. Faris, Revolutions Without Revolutionaries? Social Media Networks and Regime Response in Egypt. Ph.D. thesis, University of Pennsylvania, 2010
26. M. Fonoberova, V.A. Fonoberov, I. Mezic, J. Mezic, P. Jeffrey Brantingham, Nonlinear dynamics of crime and violence in urban settings. J. Artif. Soc. Soc. Simul. 15(1), (2012)
27. D.R. Forsyth, *Group Dynamics*, 5th edn. (Wadsworth, Cengage Learning, 2010)
28. Freedom House, Freedom in the World, Individual Country Ratings and Status. https://freedomhouse.org/report-types/freedom-world, 2015. Accessed 2015-07-13
29. J. Galtung, Violence, peace and peace research. J. Peace Res. **6**(3), 167–191 (1969)
30. J. Galtung, Cultural violence. J. Peace Res. **27**(3), 291–305 (1990)
31. N. Gilbert, *Agent-Based Models (Quantitative Applications in the Social Sciences)* (SAGE Publications, Thousand Oaks, 2007)
32. N. Gilbert, K.G. Troitzsch, *Simulation for the Social Scientist*, 2nd edn. (Open University Press, Maidenhead, 2005)
33. B. Gilley, The meaning and measure of state legitimacy: Results for 72 countries. Eur. J. Polit. Sci. **45**, 499–525 (2006)
34. B. Gilley, *The Right to Rule. How States Win and Lose Legitimacy* (Columbia University Press, New York, 2009)
35. M.S. Granovetter, Threshold models of collective behavior. Am. J. Sociol. **83**(6), 1420–1443 (1978)
36. V. Grimm, U. Bergern, D.L. DeAngelis, J. Gary Polhill, J. Giskee, S.F. Railsback, The ODD protocol: A review and first update. Ecol. Model. **221**(221), 2760–2768 (2010)
37. T.R. Gurr, Psychological factors in civil violence. World Politics **20**(2), 245–278 (1968)
38. T.R. Gurr, *Why Men Rebel* (Routledge, 2011). Fortieth Anniversary Paperback Edition
39. J.F. Hair, W.C. Black, B.J. Babin, R.E. Anderson, *Multivariate Data Analysis*, 7th edn. (Pearson, Upper Saddle River, 2009)
40. A. Ilachinsky, *Artificial War. Multiagent-Based Simulation of Combat* (World Scientific, Singapore, 2004)
41. J.-W. Kim, R.A. Hanneman, A computational model of worker protest. J. Artif. Soc. Soc. Simul. **14**(3), (January 2011)
42. T. Kuran, Sparks and prairie fires: a theory of unanticipated political revolution. Public Choice **61**, 41–74 (1989)
43. T. Kuran, Now out of never: The element of surprise in the east european revolution. World Politics **44**(1), 7–48 (October 1991)
44. C.M. Lemos, R.J. Lopes, H. Coelho, On legitimacy feedback mechanisms in agent-based models of civil violence. Int. J. Intell. Syst. **31**(2), 106–127 (February 2016)
45. C.M. Lemos, On Agent-Based Modelling of Large Scale Conflict Against a Central Authority: from Mechanisms to Complex Behaviour. Ph.D. thesis, ISCTE - University Institute of Lisbon and Faculty of Sciences of the University of Lisbon, December 2016. ISBN: 978-989-8862-62-4
46. K. Lorenz, *On Aggression* (Routledge Classics, London, 2002)

47. P. Machamer, L. Darden, C. Craver, Thinking about mechanisms. Philosophy Sci. **67**(1), 1–25 (2000)
48. M.D. Makowsky, J. Rubin, An agent-based model of centralized institutions, social network technology, and revolution. PLoS ONE **8**(11), e80380 (2013)
49. B. Milanovic, All the GINIS Dataset, 2014. Accessed 2016-03-28
50. J.H. Miller, S.L. Page, *Complex Adaptive Systems* (Princeton University Press, Princeton, 2007)
51. A. Mood, F.A. Graybill, D.C. Boes, *Introduction to the Theory of Statistics*, 3rd edn. (McGraw-Hill, New York, 1974)
52. A. Moro, Understanding the dynamics of violent political revolutions in an agent-based framework. PLoS ONE **11**(4), 1–17 (2016)
53. M.E.J. Newman, Power laws, Pareto distribution and Zipf's law. Contemp. Phys. **46**(5), 323–351 (September-October 2005)
54. North Atlantic Treaty Organization, NATO Standardization Agency, *AAP-6 NATO GLOSSARY OF TERMS AND DEFINITIONS*, 2008
55. R Core Team, *R: A Language and Environment for Statistical Computing* (R Foundation for Statistical Computing, Vienna, Austria, 2013). ISBN 3-900051-07-0
56. A. Raine, *The Anatomy of Violence: The Biological Roots of Crime* (Vintage, New York, 2014)
57. L. Ray, *Violence & Society* (SAGE Publications, London/Thousand Oaks/New Delhi/Singapore, 2011)
58. R.A. Rescorla, A.R. Wagner, *A Theory of Pavlovian Conditioning: Variations in the Effectiveness of Reinforcement and Nonreinforcement*, vol. 20, pp. 64–69. Classical Conditioning ii: Current Research and Theory edition (Appleton-Century-Crofts, New York, 1972)
59. W. Revelle, *psych: Procedures for Psychological, Psychometric, and Personality Research*, 2016
60. H. Rheingold, *Smart Mobs: The Next Social Revolution* (Basic Books, 2002)
61. R.W. Rummel, *Understanding Conflict and War Volume 2: The Conflict Helix* (SAGE Publications, Beverly Hills, 1976)
62. S.J. Russel, P. Norvig, *Artificial Intelligence. A Modern Approach* (Prentice Hall, Englewood Cliffs, 1995)
63. I. Salehyan, C. Hendrix, *Social Conflict Analysis Database Version 3.1*, November 2014
64. T.C. Schelling, Models of segregation. Am. Econ. Rev. **59**(2), 488–493 (1969)
65. T.C. Schelling, Dynamic models of segregation. J. Math. Sociol. **1**, 143–189 (1971)
66. G. Sharp, *From Dictatorship to Democracy*, 4th edn. (The Albert Einstein Institution, Boston, 2010)
67. H. Tajfel, Social identity and intergroup behavior. Soc. Sci. Inf. **13**, 65–93 (1974)
68. H. Tajfel, J. Turner, *An Integrative Theory of Intergroup Conflict*, volume *The Social Psychology of Intergroup Relations*, chapter 3, pp. 33–47 (Brooks/Cole Publishing Company, Monterey, 1979)
69. The English Indices of Deprivation 2010, UK Department for Communities and Local Government, March 2011
70. The Fund for Peace, Fragile States Index. http://fsi.fundforpeace.org/data, 2015. Accessed 2015-11-09
71. The Robert S. Strauss Center, Social Conflict Analysis Database. https://www.strausscenter.org/scad.html, 2015. Accessed 2015-07-25
72. J.C. Thiele, R Marries NetLogo: introduction to the RNetLogo package. J. Stat. Software **58**(2), 1–41 (2014)
73. J.C. Thiele, W. Kurth, V. Grimm, RNetLogo: An R package for running and exploring individual-based models implemented in NetLogo. Methods Ecol. Evol. **3**(3), 480–483 (2012)
74. Uppsala Conflict Data Program (UCDP), *UCDP/PRIO Armed Conflict Dataset Codebook, Version 4-2014a* (Centre for the Study of Civil Wars, International Peace Research Institute, Oslo (PRIO), 2014)
75. United Nations Department of Economic and Social Affairs, World Population Division, World Population Prospects, 2015 Revision. http://esa.un.org/unpd/wpp/Download/Standard/Population/, 2015. Accessed 2015-12-16

76. P.-O.H. Wikström, K.H. Treiber, Violence as situational action. Int. J. Conflict Violence **3**(1), 75–96 (2009)
77. U. Wilensky, *NetLogo*. Center for Connected Learning and Computer-Based Modelling (Northwestern University, Evanston, IL, 1999)
78. U. Wilensky, *NetLogo Rebellion Model* (Northwestern University, Evanston, IL, 2004)
79. Wolfram Research, Inc., *Mathematica*, version 10.1 edn. (Champaign, Illinois, 2015)
80. M. Wooldridge, *An Introduction to MultiAgent Systems*, 2nd edn. (Wiley, New Yor, 2009)
81. World Health Organization, Geneva, *World Report on Violence and Health: Summary*, 2002
82. H. Xia, H. Wang, Z. Xuan, Opinion dynamics: a multidisciplinary review and perspective on future research. Int. J. Knowledge Syst. Sci. **2**(4), 72–91 (2011)
83. S. Žižek, *Violence* (Profile Books, London, 2008)

FSC
www.fsc.org
MIX
Papier | Fördert gute Waldnutzung
FSC® C083411

Kolibri 360 GmbH
Ferdinand-Jühlke-Straße 7
99095 Erfurt, Deutschland
produktsicherheit@kolibri360.de